Home Insulation

Do it Yourself
& Save As Much As 40%

By Harry Yost

A Storey Publishing Book

Storey Communications, Inc.
Schoolhouse Road
Pownal, Vermont 05261

For additional information please contact
Storey Communications, Inc., Schoolhouse Road,
Pownal, Vermont 05261.

Cover and text design by Cindy McFarland
Cover art by Robin Brickman
Edited by Roger Griffith
Architectural drawings by Raymond Wood
Indexed by Kathleen D. Bagioni

Printed in the United States by The Book Press
Second Printing, April 1992

LIBRARY OF CONGRESS CATALOGING-IN-PUBLICATION DATA

Yost, Harry, 1930-
Home insulation : do it yourself and save as much as 40% / by Harry Yost.
p. cm.
Includes bibliographical references and index.
ISBN 0-88266-741-6 — ISBN 0-88266-694-0 (pbk.)
1. Insulation (Heat)—Amateurs' manuals. 2. Dwellings—Energy conservation—Amateurs' manuals. I. Title.
TH1715.Y67 1991 91-50001
693'.832—dc20 CIP

CONTENTS

Introduction

The cost of heating a home keeps rising and we begin to wonder how we will pay to keep warm. Then we are told the protective layer far above us is being destroyed by the use of fossil fuels, many of the fuels we use to heat our homes. The problem is broadcast on the nightly news. Oh yes, stations also broadcast the solutions and they all sound too expensive for the average homeowner. Must we learn to hibernate every winter like a bear?

For the homeowner willing to expend some effort and money, or for the family building a new home, there is a partial solution to both the cost of keeping warm and the problems with the environment: Insulation. Insulation? That stuff is expensive, too!

Of all the proposed solutions or partial solutions to a homeowner's energy crisis (financial crisis may be more to the point), insulation is one that pays for itself. Adding insulation to an existing house or having the maximum amount of insulation put in a new house can reduce the fuel burned enough to pay back the investment, often many times over during the life of a house.

Homeowners are very conscious about the use of insulation, but knowing about it and doing something about it are two different things. As late as the mid-1980s, U.S. government agencies estimated that as many as half of all houses in the country were underinsulated. That translates into a lot of energy being wasted and a lot of the environment being damaged to mine coal or drill for oil, not to mention the wholesale destruction of forested valleys as water is dammed to generate the electricity we use.

The problem isn't convincing homeowners to use insulation. Most want to use every method they can find to reduce heating or cooling bills — as long as it doesn't cost them anything.

Part of the problem is the attitude of builders and remodelers. The bottom line keeps getting in the way. Contractors submitting bids for construction may try to shave the insulation job a little in order to get the job and stay in business. Developers, who hope they can sell the house before the bank wants payment on the money they borrowed, will cut back on insulation rather than cut into their profits.

To add to the problem, many building codes not only don't consider insulation, but even outlaw construction methods that take advantage of insulation. No state has yet adopted a building code specifying how much insulation should be in a new house. Some do have model codes in the works, but these offer goals, not requirements. Federal agencies that insure home loans are often the only governmental organizations requiring any insulation, and they often disagree on such questions as how much.

Confusion abounds about insulation. Each manufacturer makes "better" material than everyone else. Each builder of new homes puts in the "maximum" amount of insulation. Each insulation contractor uses only the "better" brands and has the "best" method of installation.

To further muddy the situation, until recently there was no one place where a homeowner could get an unbiased comparison of the different brands and types of insulation. "Independent" testing laboratories were paid for by the industry, and their data aren't usually available to the public. Nor did any government agency put forth any data a layman could use to tell if an insulation salesman's claims were inflated. Some universities did excellent research and compiled a lot of information, but in most cases it wasn't available to the public.

Now that is changing. In the early 1970s the Canadian government established a testing laboratory and subsequently began to publish the information for public use. The U.S. government is giving grants to independent organ-izations for insulation research, and several federal agencies are compiling and publishing data of use to homeowners. Non-profit organizations such as Alaska Craftsman Home Program, Inc., are publishing data in ways that the average homeowner can understand. Insulation has even gone high-tech. Both the U.S. and Canadian government agencies have worked up programs so anyone with a PC can design an insulation package. It even enables the user to project the savings from different amounts of insulation.

With this emphasis on insulating a new house, it may seem that owners of existing houses have been forgotten. This isn't the case, though more work has been done in the area of new home construction rather than "retrofitting." (That's a word coined to define what a homeowner does with his existing house to benefit from the energy savings of insulation.)

This book is the result of some thirty-five years in and around the construction trades, information gathered from a variety of sources, and *lots* of sad experiences of myself and others. It is an attempt to address this problem. It tells how you, the homeowner, can do much of the work on an existing house; how you determine the value of insulation in a house you are considering buying, and how to judge if a builder's plans show the needed amount of insulation. Simply, it tells how to get the most for your insulation dollar.

When I was growing up in northeast Washington state, insulation was the exception rather than the rule. Even with -30° F winters, little thought was given to preventing heat loss. The response to cold weather was to pile more wood or coal into the stove. Sometimes people stuffed newspapers or rags around doors and windows to cut down the drafts. Bedrooms weren't heated unless they were above the heated living space. The response to cold nights was to put more blankets on the beds.

My first encounter with insulation was to see how it kept things cold, rather than warm. Ice from a nearby lake was stored in a large ice house in our town. The building was insulated with sawdust and wood shavings. A few of the homeowners concluded that if they used the same material they could keep their houses cooler in summer and warmer in winter. They filled the walls and attics with sawdust and shavings, which was free for the hauling from the local sawmill.

Insulation materials were on the market before the 1930s. One was what we now call cellulose fiber, made of ground newspapers and borax powder. The process was patented in 1915 under the trade name Sci-Pro-Cel. Cork was also used in ice boxes and some larger freezers. Fiberglass was used in refrigerators beginning in the 1920s.

Just prior to World War II, a company about fifteen miles from where I lived began manufacturing an insulation made from wood fibers and a by-product of the refining of a steel-toughening mineral called magnesite, which was mined locally. The insulation was called Thermite, as I remember. It was made in large sheets about 2 inches thick and was used for walk-in freezers, among other things. The company shut down just after the war.

In the early 1950s, I was living in southern California and purchasing my first house. I became concerned about the heating system, which consisted of a free-standing, unvented gas heater and a fireplace with a perforated gas pipe to ignite the wood. Subsequent to installing a small forced-air gas wall furnace, I became concerned about getting the most heat for my energy dollar. Gas bills ran $8 to $12 a month, with gas used for heating water and cooking.

I installed 1½ inches of fiberglass insulation in the attic, which was the only place any could be put because the walls were only 2¼ inches thick. We were well satisfied with the savings, which I guess proves any insulation is better than none.

Later on I lived in Washington, then western Oregon, and finally Alaska. As I learned more about savings generated by the use of insulation, I put more in the houses I owned, and became more concerned about how much was already there when I purchased a new house. (Not enough, I know now but didn't know then.)

In Alaska I worked for one of the departments of the University of Alaska and did remodeling work on older houses in the area. Part of my duties was to do maintenance work on houses owned by the university. These had been constructed by the U.S. Department of Agriculture between 1917 and the mid-1950s.

The newer houses were insulated with 1½ inches of a crepe paper material in both walls and ceilings. The older houses had a layer of dust. All were heated with oil. In the energy crunch of the 1970s, we scrambled to find some way to add insulation to these structures. Blowing 6 inches of cellulose fiber insulation into the attics of the houses made a dramatic difference in both fuel consumption and comfort.

Late in the 1960s engineers in our department completed a five-year study on the use of various types of insulation in that climate. We used what they learned as we retrofitted some of the older houses. Some of this was pretty crude, but it worked. We hand-poured mineral insulation into the walls, for instance. Fuel savings were very noticeable. Among things determined by that exper-iment was that many manufacturers' claims about the resistance to heat loss of their materials were somewhat overstated.

Even with free data and demonstrated savings in energy bills, many builders considered insulation a nuisance. It was something put in to make the lender happy so the checks would keep rolling in.

In the late 70s and early 80s when the pipeline and oil boom was at its height, builders form all over the Lower 48 converged on Alaska to take advantage of the easy money to be made building houses and commercial buildings. Some came from areas where insulation had never been considered, and inadequately insulated houses were only some of the problems left in their wake. Alaska, along with most other states, didn't have codes to force builders to put in proper insulation or vapor barriers, so it was "whatever the traffic will bear."

With the Alaskan crash of the late 1980s and the subsequent repossession of large numbers of houses by lending institutions, proper insulation became a critical item. Many houses remain unsold, even with rock-bottom prices (half or less than the original price) because they are not adequately insulated. With electricity and fuel oil so expensive, home buyers are getting very selective about how they spend their hard-earned dollars.

I have also done construction and remodeling work in Wisconsin, Montana, and Oklahoma, where insulation is used as much for keeping cool as it is for keeping warm. In Oklahoma and Wisconsin the summers are hot and humid, with winters cold and humid. Montana summers are hot and dry, with cold, dry winters. Insulation, along with plenty of attic ventilation and radiant barriers, must be installed by homeowners to get the most value for their energy dollar. In the buildings I worked on in Oklahoma, the annual cost of cooling was nearly double the cost of heating, using the same heat pump system for both.

Despite evidence that insulation worked as well for cooling as for heating, builders in Oklahoma skimped on insulation. They favored attic ventilation, installing powered systems in many houses and commercial buildings.

In all of the places I've worked, I've found that people are convinced of the value of insulation. But they want details. What is the best way to install it? How much should be used? And if I use enough of it and install it properly, how much money will I save?

The aim of this book is to provide some of those answers. It will, we hope, lead to a reduction in our use of fossil fuels, and in that way contribute to the solution of our environmental problems, both personal and national.

THE BASICS

1. All About Insulation

THE WIDE ARRAY of insulations on the market today can confuse the potential user. Costs range from a few cents to over a dollar a square foot. And high cost doesn't necessarily mean high efficiency. While all insulation eventually pays for itself, some of the more expensive aren't any better than those costing half as much.

Here is a description of several insulations and the forms they are available in, along with some estimated costs for using them.

Fiberglass

Also called glass wool or spun glass, fiberglass is probably the most versatile insulation on the market today. It is used in buildings, hot water heaters, cars, refrigerators, and wherever there is a need to insulate from heat, cold, or sound.

A major reason for such wide use is its low cost. It probably gives more insulation value for the money than any other kind available. In most sections of the United States, 6 inches thick fiberglass can be purchased for thirty to forty cents a square foot.

Glass Fibers

Fiberglass, as the name implies, is made of glass fibers or thin threads. It looks and feels like colored wool, hence the name glass wool used in some parts of the country. It is white, yellow, or pink. The pink has been mistakenly eaten by small children because it looks like cotton candy, so be careful with scraps.

It is available in batts, rolls, and loose and rigid forms. The batts come in thicknesses from 1 to 8 inches and widths to fit between 16- or 24-inch studding or joists. Batts are either 48 or 96 inches in length.

Fiberglass rolls are made in 1½-, 3½-, and 6-inch thicknesses. Both rolls and batts come with no backing, a backing of aluminum foil, an aluminum foil-tar-kraft paper combination, kraft paper, or a cloth-like vinyl material.

An unbacked roll type 1 inch thick and either 4 or 6 inches wide is also available for use as sill seal (between the foundation and mud sill or between the bottom plate and the floor under outside walls) or as pipe wrap. For commercial purposes, fiberglass also comes in rolls 3½ inches thick and either 36 or 72 inches in width and either vinyl or aluminum foil backed.

Fiberglass insulation won't burn, but the backings will. It should not be put in direct contact with chimneys or recessed light fixtures as it may transmit heat to other flammable materials.

Rigid Fiberglass

Rigid fiberglass insulation has been treated to make the fibers stiff and thus maintain its thickness. It is generally yellow and backed with either aluminum foil or a thin vinyl. The foil-backed is used to insulate rectangular heating or cooling ducts and sometimes as sub-siding on houses.

Vinyl-covered rigid fiberglass is mostly used in suspended ceilings as drop-in panels. The vinyl finish is a crinkly off-white and gives a neat, clean look. For use in suspended ceilings it is often precut to 24 by 24 inches or 24 by 48 inches, 1 inch thick. It is also available in sheets 4 feet wide and up to 12 feet long, in thicknesses from 1 to 4 inches.

Loose (blown) fiberglass is used for horizontal and closed applications such as attics. It is installed with a machine using air pressure to blow it into place. It comes tightly packed in large plastic bags. Usually a chart on the bag gives the coverage per bag at different thicknesses. It has a lower insulation value per inch than the batt/roll or rigid types, but this is offset by a lower cost per square foot. Experienced insulators should apply this, as the pressure of the air and speed of the blower affect how the material goes into place.

All fiberglass insulations are classified as "verminproof," but mice will make nests and tunnels in it.

Polystyrene

This material is often called Styrofoam, which is a registered trade name used by one manufacturer. It is available in rigid sheets, 4 by 8 feet in size and ½ to 4 inches thick. It can be purchased from some manufacturers in blocks as large as 16 inches thick, 4 feet wide, and 16 feet long for such uses as docks or swimming floats. It is made of bubbles of plastic, expanded under pressure. It may be white, white with blue or green specks or tint, or solid blue in color, depending on the manufacturer, with the colored material having a higher insulation value than the pure white. However, the difference in the cost for the better quality usually isn't offset by the slight increase in insulation value.

Polystyrene insulation can be used for sub-siding, glued to concrete basement and foundation walls, both inside and outside, placed under concrete floors and footings, or cut to fit between framing members. It is waterproof, and is used as a flotation material in many small boats. However, it is flammable when exposed to fire, and the fumes from burning are toxic.

A unique type available in some localities is a hollow block, 16 inches high, 8 inches thick, and 48 inches long, called a Foamform. The blocks are tongue-and-grooved to interlock. When placed on a concrete footing and filled with concrete, with reinforcing rods placed vertically and horizontally as directed, they make a foundation or basement wall insulated on both sides. They lend themselves to do-it-yourself work as the concrete must be poured slowly and carefully in small amounts, and the pouring may be stopped and started as the worker desires.

The cost of polystyrene varies widely, depending on the manufacturer and how far you are from the factory.

Polyurethane

It is usually called urethane. It comes either in 4 by 8-foot sheets or as a foam which can be applied to any shape or angle surface with special equipment. It is also sold in small aerosol cans with applicators designed to put it in small spaces such as around door or window frames where it dries and seals against air leaks. The sheets are either white or yellow and have a sandy texture. The foam is yellowish-white and has a sandy texture after it dries.

Soon after it came on the market, urethane developed a bad reputation because it burned readily and emitted a toxic smoke. This problem has been allayed somewhat by the addition of fire-retardent chemicals, but some building codes prohibit its use unless it can be protected from fire.

Waterproof

Polyurethane is waterproof and either sheets or foam can be used on the outside of basement or foundation walls, buried tanks, earth houses, or roofs. If it is exposed to sunlight it must be coated with a special paint to keep it from degrading (evaporating).

When first introduced it was sometimes called "the miracle insulation" because the foamed type had the highest insulation value of any insulation on the market. Canadian government tests then found that it lost insulation value over time, and that within five years of installation it had no more insulation value than polystyrene. The cost was considerably more, though.

The cost of the rigid type varies, depending on shipment costs. The foam type can cost anywhere from twenty-five to seventy-five cents a square foot, 1 inch thick.

Urea-Formaldehyde

This is a foam insulation which was used extensively in older houses that had no insulation in the outside walls. It was applied through holes drilled in the siding, one at the bottom and one at the top, between each pair of studs. The foam was forced into the bottom hole until it came out of the top one, to fill the space with insulation.

Its use has been generally discontinued because of the effect the formaldehyde has on some people. However, because it was used extensively in some areas, many houses on the market have it in the walls. If contemplating buying an older house that was insulated with urea-formaldehyde, you should test to see if there is any formaldehyde residue in the building, and if it will bother any future residents. Recent Environmental Protection Agency tests indicate the formaldehyde levels drop over time. Removing urea-formaldehyde insulation is costly. The interior sheathing must be removed and the material carved out of the walls.

Urea-Tripolymer

This foam material is similar to urea-formaldehyde without the formaldehyde, and is applied the same way. It is waterproof and has a low fire hazard rating. It can shrink when it dries and may leave gaps around the edges. Costs vary according to location.

Isocyanurate

This has the highest insulation value of any material available to the homeowner; its insulation value is double that of fiberglass. It is a rigid material available in sheets 1 to 2 inches thick and 2 by 8 feet or 4 by 8 feet in size. It may have aluminum foil on one or both sides, depending on the manufacturer. It can be glued to basement or foundation walls, but is best suited for use where other insulations cannot be put in thick enough to gain the desired insulation value. It is expensive, but if needed for a special place can be worth the extra cost.

It is flammable and gives off toxic gases when burning. The cost varies, depending on the distance it is shipped.

Mineral Wool

Most often called rock wool, mineral wool looks like dirty wool and is dusty when handled. It is generally used for horizontal applications and is placed by blowing it with a machine or pouring it by hand. It will settle over time, lowering its original insulation value. It is heavier than other types of insulation and will soak up moisture. It cakes after being wet and loses most of its insulation value.

It is also sold in envelope-type batts, enclosed by a kraft paper material. It was very popular in the 50s and early 60s in some parts of the country and can be found in houses from that time. When buying a house insulated with mineral wool, plan to add either blown fiberglass or cellulose fiber to bring the insulation in the attic up to recommended levels. Replacing it or adding more insulation to the walls may also be considered.

Cellulose Fiber

This inexpensive insulation is made from recycled material, mostly ground newspapers mixed with borax or aluminum sulfite powders. The process was patented in 1915 under the name "Sci-Pro-Cel", which is still used in some localities. Because of the ease of manufacturing it (all that's needed is a farm hammermill and a source of old newspapers), it is made by many small companies around the country and the costs and quality vary considerably.

The borax and aluminum sulfite are mixed with it to make it fire resistant, and also make it slide through blower hoses easily. It is mostly used for horizontal applications such as attics, but some makers mix a water-activated glue with it, then with special equipment apply it to walls. Several days' drying time is necessary before any other work can be done on the walls.

It has a slightly higher insulation value than blown fiberglass and costs less. It is not considered verminproof, and can soak up moisture and degrade considerably if allowed to get wet. It also mats and loses insulation value if wet. Though not generally considered flammable, it must be kept away from chimneys, range hood vent pipes, and recessed lights. Many rental companies rent small blowers to homeowners to apply this material, and it can be easily placed by an adept do-it-yourselfer with a little care.

A problem has surfaced with the use of aluminum sulfite powder. In a house where the roof is supported by factory-built trusses, and with poor attic ventilation and/or no vapor barrier, accumulated moisture can act with the aluminum sulfite to attack the galvanized gang nailers on the trusses and dissolve them, weakening the trusses and the roof. If there is any question whether the powder used was aluminum sulfite, this should be ascertained before the material is used, and steps should be taken to prevent trouble. One essential step is to have plenty of attic ventilation.

Expanded Mica

This product is made from the mineral mica. It may be called vermiculite or perlite, which are similar and are sometimes used for insulating, as well as for bedding material in greenhouses.

The material is used mostly for horizontal installations in small spaces as it can be poured by hand easily. Where there is access, it may also be poured into the walls of older houses, but care must be taken to compensate for any settling that will occur. (See section on insulating older houses.)

Expanded mica has a lower insulation value than fiberglass. Cost will depend on locality as the material is mined in only a few places in the United States and Canada. It is shipped in large paper or plastic bags, and shipping is expensive.

Cellulose Fiberboard

This is made of ground wood fibers pressed into a soft, rigid board.

Among other names, it is called Celotex, Firtex, Fiberboard, Pressboard, and black board when impregnated with tar. The tarred version is often used as sub-siding on houses.

Pieces 4 by 8 feet and ½ inch thick are often mounted in frames and used for bulletin boards by schools and businesses as it is easy to tack up items on it. This particular type is painted white on one side. In this version it is also used for ceiling and wallboard on vacation homes or other light-duty buildings. Some of the less expensive ceiling tiles are made of it as well as the less expensive panels for suspending ceilings.

In the past, it has been manufactured in panels 2 by 4 feet in size and up to 2 inches thick to be used for insulation in walk-in freezers or cold storage lockers.

It is flammable and will soak up moisture. Ceiling tiles will deform and discolor and must be replaced if they get wet.

2. Estimating Your Needs

BECAUSE DIFFERENT INSULATIONS have different insulation values per inch of thickness, the industry has devised a simple method of comparison.

Insulations are rated by their R-values. R-value is a numerical rating and refers to a material's resistance to heat loss. The higher the R-value, the greater the material's resistance to heat loss. Nearly all building materials have an R-value. In this chapter a chart gives the R-value for most insulations and such items as wood, concrete, and metal siding.

Using this chart, a homeowner can quickly determine how one insulation stacks up against another. Figuring the R-value against the cost per square foot will show that some insulations, despite a higher R-value per inch, are more costly than a type with a lower R-value per inch. As mentioned in the previous chapter, fiberglass, though having a lower R-value per inch, could be much less costly, even with heavier building framing, than isocyanurate, which has double the R-value.

R-value is determined by engineers from measurements made to determine the U-value. U-value represents the total heat transmission (loss) in Btu's (British thermal units) per square foot, per hour, with a one-degree temperature difference between the inside and the outside. The *lower* the U-value, the better the product will insulate.

Use of R-Values

Knowing the insulation value of the different components of a house is necessary to determine how well insulated the building is. Having a high

R-Values of Common Insulation and Building Materials

(Per inch of thickness unless otherwise noted)

MATERIAL	R-VALUE
Fiberglass (batt or roll)	3.7
Fiberglass (rigid)	3
Fiberglass (loose)	2.2 to 2.7
Cellulose fiber (loose or sprayed)	3.7
Expanded mica	3.1
Polystyrene	3.5 to 4.2
Polyurethane (foam or boards)	4 to 6
(Foam may be as high as 9 when installed, but loses value for first five years before stabilizing at 4 to 6.)	
Isocyanurate	7 to 8
Mineral wool	3
Windows (single pane)	0.04
Windows (double glazed with ¼-inch sealed air space)	0.79
Windows (double glazed with ½-inch sealed air space)	0.97
RDG (single pane with removable storm window)	0.69 (average)
(Tinted, insulated and otherwise treated windows will vary from the above, according to manufacturer. Triple-glazed windows will also vary, according to manufacturer.)	

COMMON BUILDING MATERIALS	R-VALUE
½-inch cellulose fiberboard	0.6
½-inch gypsum board (Sheetrock)	0.45
Metal siding	0.00
Plaster	0.16
Plywood (1 inch thick)	1.0
Hardwoods (at 5 percent moisture content, 1 inch thick)	0.9
Softwoods (at 5 percent moisture content, 1 inch thick)	1.25
Asphalt roofing (per thickness) shingles or roll	0.15
Hollow concrete blocks (8 by 8 by 16 inches)	1.0
Poured concrete (per inch thickness)	0.08 to 0.09
Carpet on pad	2.08

(Most of these figures are the result of a five-year test of insulations done in the late 1960s, from publications of the Cooperative Extension Service, University of Alaska.)

R-value insulation in part of the building doesn't mean the building is well insulated, only that a portion of it is. For instance, a house with R-19 insulation in the walls may have large picture windows in those same walls, the glass of which may only be R-0.97. In addition, there is a lot of wood in a wall, and wood also has a lower R-value than insulation.

There is also another compelling reason for knowing the average R-value of the various parts of a house being purchased. When lending money for buying a home, some lenders want to know if the house has a given amount of insulation in the various parts. Some, particularly those insured or funded by the government, require a minimum *average* R-value in the different sections of a house.

There are more than thirty U.S. government loan or loan insurance programs. Most people who borrow money for a house will be involved, directly or indirectly, with at least one of these agencies. If a house is underinsulated, some agencies may lend additional money to bring it up to standards, or may know of government programs that will pay for the insulation upgrade.

Figuring R-Values

Since each component of a house has an R-value, and that value is known (see the preceding chart), it is relatively easy to find the average R-value of a given part of a house, using simple arithmetic.

The example that follows shows how to figure the average R-value for a single wall, 20 feet long and 8 feet tall, framed with 2x4 lumber. It has a solid core wood door 6 feet 8 inches tall, 36 inches wide and 1¾ inches thick, and a double-glazed window 6 feet wide by 4 feet 6 inches tall. The exterior is wood lap siding nominally ½-inch thick over cellulose fiberboard the same thickness. The interior finish is ½-inch gypsum board (Sheetrock) and the insulation is 3½-inch fiberglass.

Finished 2x4 framing lumber is actually only 1½ by 3½ inches. In standard framing, a carpenter figures one stud per lineal foot of the wall, so there should be twenty studs in our example wall.

To get the number of square inches of wood in the wall, multiply the facing width of the studs (1.5 inches) by the number of studs (20) by the height of the studs (8 feet or 96 inches). 1.5 x 20 x 96 = 2880 square inches.

Divide this by the number of square inches in a square foot (144) and you get an answer of 20 square feet of solid wood.

Now you can add together all of the non-insulated areas. They are:

Solid wood	20 square feet
Window	27 square feet
Door	19 square feet
Total	66 square feet

To find the square footage of the insulated area:

Total wall area	160 square feet
Uninsulated area	- 66 square feet
Insulated area	94 square feet

Now you can determine the R-value of each part of the wall, using the table:

½-inch wood siding	R-0.625
½-inch cellulose fiberboard	R-0.6
½-inch Gypsum board	R-0.45
3½-inch solid softwood framing	R-4.375
Double glazed window, ½-inch airspace	R-0.97
Solid wood door, 1¾-inch thick	R-2.1875

Add the R-values through both the solid wood sections and the insulated sections:

Total R-value of solid wood section	6.05
Total R-value of insulated section	12.675

Multiply the square footage of each section with the total R-value of that section of the wall:

Insulated area	94 sq. ft. x	12.675	= 1191.45
Solid wood area	20 sq. ft. x	6.05	= 121
Window area	27 sq. ft. x	0.97	= 26.19
Door area	19 sq. ft. x	2.1875	= 41.5625
Total			1380.2

Divide that total by the number of square feet of the wall:

1380.2/160 = 8.63 which is the *average* R-value for that wall.

Building engineers will have a slightly higher figure, as they figure in air films and reflective surfaces. However, for the homeowner's purposes, the above will be adequate.

3. Avoiding Radon Problems

RADON IS A RADIOACTIVE GAS occurring naturally in nature. It is odorless, tasteless, and colorless and is found in varying densities in homes throughout the United States, Canada, and the rest of the world.

Because it is radioactive, it is considered harmful. How harmful is the subject of debate among researchers. Some medical scientists estimate that from 5000 to 20,000 of the deaths attributed to lung cancer each year in the United States are a direct result of exposure to radon.

These figures were derived from a study done in the 1940s and early 1950s on miners, mostly those working in uranium mines. The data gained were then extrapolated to the general population. Because radon is radioactive, it tends to amplify the damage done by other forms of pollutants found in homes, such as cigarette smoke. The study pointed out that most of the miners smoked.

Two Ways to Enter

Radon usually enters a house indirectly through the water used in the house, or directly from the ground the house is built on. Radon in water can be removed easily in two ways, with a filter or by aerating the water in a chamber and releasing the radon into the outside atmosphere.

There are also two ways of dealing with radon threatening a house from the ground: prevention and removal. The easiest way for the homeowner is prevention. By placing an impervious cover such as 10 mil (.010 inch thick) polyethylene film, called a vapor barrier, on the ground under a house, a homeowner can prevent the transmission of most of the radon from the soil into the air of the house.

Use a Sealant

In a house with a basement, all cracks, spaces around pipes, and any other points where the gas could enter must be sealed with a non-hardening sealant such as a silicone-based caulking compound. There can be no slits, cracks, or other openings, no matter how small, in the material used.

Experiments have been conducted using polyurethane foam sprayed in a thin layer on the ground under a house. That is more expensive and hasn't been totally satisfactory as the polyurethane hardens and can be cracked by someone walking on it, or it can shrink away from foundation walls.

Air Exchange Unit

The other method of dealing with radon is ventilation. This can be simple, such as having vents in the crawl space or opening a couple of windows. It can also be more complicated and expensive, such as using an electrical device called an air exchange unit.

An air exchange unit insures that a given amount of air is exhausted and replaced in the house every hour.

The warm air that is exhausted is used to warm or temper the incoming cold air, so the heat from the house isn't wasted. Using an air exchange system is the best way to keep a low level of radon in a house and reduce the amount of any other pollutants that may be present.

The drawbacks of this system are that it uses a constant amount of electricity to operate and increases heating costs because it is not 100 percent efficient in heating the incoming air. However, compared with leaving windows open in cold weather, it is less costly. It is also a machine, so there may be mechanical problems at times.

Threat of Higher Level

The homeowners' dilemma is that in attempting to make their houses more energy efficient, they will make them tighter and therefore increase the amount of radon and other pollutants to unsafe levels. In order to identify cold drafts, some have gone so far as to have their houses pressurized, then detected and sealed all the air leaks, thus making them virtually airtight. This quickly increases the level of radon, if any is present locally, other pollutants, and moisture inside the house. In order to control those, some tradeoffs must be made, and an air exchange system is about the only way to go.

The only way homeowners can find if their houses have an excessive amount of radon is to test for it.

Testing for Radon

Radon testing has become one of the newest entrepreneurial devices for separating homeowners from their money. No simple way is available which homeowners may use to determine the radon concentration in their houses. They must either purchase a test kit, available at many building supply and home improvement centers, or hire a tester to check the house.

A kit must be left in the house for from twelve hours to several months and then sent to a laboratory for examination. Prices for kits range from $10 upward, and the price may not include the cost of the laboratory examination. (Here in Alaska, a kit is being sold by a building supply house. It costs $19, including the cost of the lab work. It is left in the house for twelve hours.)

Testing a house by a firm specializing in that work can cost several hundred dollars. This is more inclusive, and often takes several months. Most often, the heaviest concentration is in the winter, when there is less air circulation into the house. Thus, ideally, tests should be made throughout one year.

In some sections of the country, state and federal agencies have established testing programs. Ask about this through the state's Radiation Protection Office, the nearest Environmental Protection Agency office, or the nearest University Extension Service office. If none of these agencies does any testing, one of them should have a list of reputable local firms which do.

A simple way to find if you live in a problem area is to contact the nearest EPA (Environmental Protection Agency) office. The EPA has conducted extensive tests in an attempt to pinpoint problem areas in the country.

Even this information could be misleading. House-by-house testing has been done in some high concentration areas with surprising results, showing widely varying amounts from house to house. Even if your neighborhood has a high concentration, your house may not be as bad.

Cross-Purposes

As with other items on the list of energy conservation, radon removal and prevention sometimes work at cross-purposes to other things that are done or recommended to be done. For instance, a simple way to reduce radon in a house with a crawl space is to have good ventilation there with large spaced vents in the top of the foundation or the rim joist of the house.

In colder areas, this then exposes water or sewer pipes to the possibility of freezing. If the vents are closed during the winter, the homeowner risks higher radon concentrations. Having the vents open in winter also lowers the temperature of the floor to uncomfortable levels. It may be necessary to insulate the floor and any pipes in the crawl space in order to use the less expensive methods of dispersing radon.

Vapor Barrier Cheapest

If placing a vapor barrier on the floor of a crawl space reduces the radon to a safe level, the homeowner should opt for that, as it is the least expensive way to deal with a cold crawl space. Foundation walls can also be insulated, with the vapor barrier extending unbroken from the floor of the house completely across the area. Again, you should test during the winter to ascertain if the work has reduced the radon in the crawl space and the rest of the house.

Protection in Basement

For a house with a basement, more work is involved. You must seal all cracks and holes in the basement floor and walls with a non-hardening caulking material. This includes the joint where the walls meet the floor and around all pipes entering through the walls and floor. Make sure the drain traps in the floor are kept filled with water. Use a non-hardening caulking such as silicone or a silicone-based material.

Basements often contain water heaters, furnaces, and clothes dryers which use quantities of air and require exhausts to the outside. These must be supplied with some form of air intake, so they don't cause a vacuum (negative pressure) in the basement and draw radon in from the surrounding soil.

The air intake should be well above ground level to keep radon in the immediate area from being inadvertently drawn into the basement. A properly designed forced air heating system should have a fresh air intake as part of the return side of the system. However, gas- and oil-fired appliances require combustion air, and if it is not supplied through a direct duct to the outside, they will draw that air from the inside of the house. If the house is tight, with all air leaks sealed as should be done in a proper energy efficient retrofit, these fired units will not operate efficiently and could be dangerous to the occupants of the house.

Flushing System

For the builder of a new house, prevention is the best way to go. Placing heavy polyethylene film under the floor of a basement or under a layer of soil under a crawl space will be necessary. (Polyethylene and insulation under a basement floor is necessary for an energy-efficient building.)

If there is a heavy concentration of radon in your area, consider the installation of a radon flushing system.

A radon flushing system consists of a thick layer of gravel, with properly labeled pipes leading into and away from it. This is topped by a sheet of polyethylene, then the concrete floor. Several times a year air, under low pressure, is forced into the mass of gravel through one pipe, and the radon-laden air is exhausted through the other pipe(s). While building this, seal any possible leaks into the basement.

The usual method of constructing a basement is to excavate, then pour footings, (leveled concrete bases around the perimeter of the building, usually 16 inches wide and 8 to 12 inches deep). When the footings are hard, you pour or build walls on the footings. Then, after the entire building has been erected, the basement floor is poured. This method of construction leaves possible leaks between the walls and the footings and between the walls, footings, and floor. Placing a non-hardening sealant such as plastic roofing compound in these areas before each additional pour of concrete goes a long way to prevent any leaks, either of radon or water.

Once the basement walls are dry, you should seal the outside with plastic roofing compound and cover it with polyethylene film. Insulation can be placed at that time. The joints between the floor and the walls should be sealed before floor covering is put into the basement.

Test First

Don't be panicked into spending a pile of money on some sort of radon reduction equipment until testing shows you have a problem. Even then there are low-cost ways radon can be eliminated or stopped from entering a house.

4. Planning Insulation Projects

TO DO ANY JOB RIGHT, you must plan it in detail before starting the work. Home construction plans may range from simple line drawings on scratch paper to elaborate, multi-page blueprints done by an architect's draftsman, but they are required to give proper direction to the work.

Begin planning for insulation as soon as a floor plan for the building is chosen. In most areas, insulation will surround the entire living space of the house. It's sometimes called an "insulation envelope." This envelope is not just something that is stuffed into the open spaces of a house after framing, exterior sub-siding, and roof are in place. In order to build the frame of the house with enough space to accommodate the required insulation, the builder must first know what the insulation value for each part of the envelope is, then decide on a framing technique that will allow that much insulation to be put in. The builder must also know what kinds of insulation will be used, as this dictates such things as the thickness of the walls and the design of the roof.

If nothing else, the builder may be required by the local building code or the mortgage lender to place a certain amount of insulation in each part of the house. Even with such requirements, it should not be assumed this will automatically be done. The amounts and placing of insulation should be clear on the blueprints and on the specification sheets. Everyone concerned with the project should understand the locations and amounts before construction begins. Otherwise, some small areas which can cause problems for years to come will be missed.

Some of these small trouble spots are in the outside corners, where inside walls join the outside walls, and where the outside walls are set up on the floor. These small areas may be blocked off by parts of the house frame, or may be ignored by the framing crew, which is usually working on a bid basis. A crew in a hurry tends to ignore these small spots, but they can cause cold areas and even frost buildup for the life of the house. Part of planning is making sure all building crews fully understand what is required of them, then checking to see that it is done as the work progresses. People being people, don't assume everything will be done the way you want just because you mentioned it.

Building codes and loan requirements aside, the strongest incentive to making sure you have a good insulation job is that you will have to live with it for some time to come, and you will have to pay the fuel bills and listen to complaints about cold and drafty places.

What Is "Code"?

The head building inspector of one of our western cities gave me some excellent advice about building codes several years ago. He commented on the often-repeated statements by builders and realtors who proudly stated their houses are "built according to code." The building inspector bluntly stated that while the phrase "built according to code" might imply an especially good building job, or one where extra attention was given to construction details, what it actually means is they did the job the cheapest way they could legally get away with. Building codes are minimum requirements. In the context of an insulation job, it may satisfy the lender, but not the people who have to live in it and pay the energy bills.

A good insulation job doesn't just happen. Insulation must be considered at each step of construction or you may find access has been blocked to some small but important area, or, if the placement of insulation has been bypassed during framing, it will take extra labor (translated as extra money) to put the proper amount of insulation in a particular place.

Once a floor plan has been decided on, the next step is to decide how much insulation (R-value) is needed where. When that is known, the construction specifications can be decided upon. Virtually all other construction specifications depend on knowing how much and what type of insulation will be used. The thickness of the walls, roof truss construction (or roof framing), and even how the foundation is to be built depend on knowing in advance how much of what kind of insulation will be used.

To illustrate: If you decide on a given wall thickness, say 6 inches before figuring out how much insulation you need, you may have to either cut back on insulation value, or spend more money for a higher R-value per inch of insulation. If you are locked into a 6-inch wall thickness, and your climate needs R-33 insulation, you will have to make some adjustments in the insulation. If, instead, you plan the walls for R-33 insulation, then you can adjust the wall specifications to take advantage of the least expensive insulation.

To bring this into a closer perspective, do you try to use 9 inches of fiberglass to attain the R-33 insulation value or 5 inches of isocyanurate to fit the 6-inch wall? Locally at this time, 9 inches of fiberglass insulation cost 60 cents a square foot, while 5 inches of isocyanurate cost $2.59 a square foot. Does the difference in the cost of framing a 9-inch wall justify the expenditure of nearly $2 a square foot more for insulation? Construction cost is something that must be considered, and is why the amount and kind of insulation must be decided upon prior to doing any structural design.

Know Code Requirements

To begin planning for insulation, you need to know what, if any, insulation is required by your local building code and your lender. These figures will probably not be the same, and in most cases will be the absolute minimum for your climate. Be sure you understand whether the figures are for the R-value of the insulation itself or are for an *average* R-value for the wall, ceiling, and other areas. As mentioned earlier, there is a vast difference between the average R-value for a wall and the R-value of the insulation in that wall.

Local energy utilities often make insulation recommendations. These may differ greatly from the ideal, even when put forth by utility engineers. In some instances, utility companies have teamed up with insulation companies to put out what they believe the insulation requirements for a given area should be. There are some problems with this. It puts the utility company in the position of recommending a particular brand or type of insulation. Also, insulation manufacturers have been known to overstate the R-value of some insulations. There is also the mistaken belief of some utility engineers, builders, and even homeowners that heat derived from different sources of energy requires different amounts of insulation to get the most out of each energy dollar.

Figures Questioned

An example of this occurred in Alaska in the mid 1970s. The local electric utility handed out *Alaska Insulation Specifications* brochures supplied by a national insulation manufacturer, essentially endorsing its product as best for electrically heated homes.

Even at the time, local building engineers knew the suggested figures were low, and research, also done locally, showed some of the insulation value figures for the fiberglass were overstated.

For the "best" insulation job, the figures were R-19 for the walls and R-34, made up of two layers of fiberglass, an R-20 and an R-14 material for the ceiling. A "good" insulation job called for R-14 (using the same material) in the walls and R-24 in the ceiling. One problem was that the R-14 material was 3⅝ inches thick. For a wall, this was to go into a space only 3½ inches wide. A quick look at the R-value chart in this book will show that University of Alaska engineers found an R-value of only 11 for a wall of that thickness. The R-24 insulation was listed as being 6¼ inches thick, which the chart shows as being slightly more than R-22 in value. These materials were not readily available, either. Cramming thicker insulation into a smaller space doesn't result in the R-value of the thicker material either. It may even result in a reduction over material designed for that space. More on this later.

By contrast, the more recent model insulation code for the same region calls for R-50 insulation in the ceiling and R-33 in the walls for the "best" job. A "good" insulation job calls for R-38 in the ceilings and R-25 for the walls. It *does not* specify a particular type, kind, or brand of insulation to attain those figures. That is left to the discretion of the builder.

Confusing?

How does the prospective homeowner cope with this welter of confusing and sometimes misleading information? One way is to hire an engineer to do the house design. This is easy and expensive. Not so easy, but definitely less expensive is to get all the information you can from various government and private agencies.

A good starting place is your local university Extension office. Many states now have model insulation codes that show figures for the good and best amounts of insulation for the different climatic regions in that state, and Extension offices should have these or be able to tell you where you can find them.

Computer Programs

If you have a personal computer, several software programs are available to enable you to develop your own insulation criteria. HOT 2000 from Canada and ZIP 1.0 developed by the National Bureau of Standards in the United States are low-cost programs designed to help the homeowner plan an insulation package. The HOT 2000 program can also project energy cost figures for different types of heating systems and fuels. (Addresses in the appendix.)

Figuring Your Needs

A couple of telephone calls and some simple arithmetic will get you the information needed to build a well-insulated house.

First, get the number of heating degree days in your locale from the Extension Service office, the local National Weather Service office, or your local utility engineer's offices. In areas where summer cooling is a consideration, also get the number of "cooling degree days."

The method of obtaining the heating degree days for each day of the year is to subtract the average temperature for that day from 65°F. Thus, if the average temperature for a day is 30°F, the number of degree days for that date is thirty-five. (65-30=35.) Adding the figures for all the days of the year when the average temperature was under 65°F gives the total heating degree days for that year in the locality where the temperature recording was done.

Cooling Degree Days

To compute cooling degree days, the same method is used, except that the base temperature (usually 70°F) is subtracted from the average temperature *over* that temperature. Thus, if the average temperature for a given day is 80°F, there would be ten cooling degree days for that date.

In climatic regions where keeping cool is as much a consideration as keeping warm, the total number of heating degree days and cooling degree days are added together for the figures needed to compute how much insulation is needed in a given area.

The first computation will be to find out the R-value of the ceiling insulation in a house. This is done by multiplying the number of degree days by .004. (10,000 degree days x .004 = 40, or R-40 insulation in the ceiling.) To compute the R-values for the rest of the house, consider the ceiling insulation R-value, 40 in this case, as 100 percent. The walls will be approximately 65 percent of 40, or R- 26. Foundation insulation should be approximately 38 percent of 40; floors, if insulated, approximately 76 percent of 40. Concrete slab floors should be around 40 percent of 40. These figures should be close to those given in most model insulation codes. They will probably be higher than called for in the local building code or your lender's specifications.

The formula "total degree days x .004" was worked out by Bob Roggasch, an Alaska builder, in conjunction with engineers from the University of Alaska's Extension Service. The formula works for the entire United States and Canada as well as for the different climatic regions of Alaska.

Use R-Value Chart

Once you have computed the R-value for each part of the house, turn to the chart of R-values (Page 12) and decide what insulation you will use in each place. You may decide to use only one kind in the whole house, and frame accordingly. If the cost of framing seems high, perhaps using a higher R-value insulation in some sections will be less expensive overall.

For instance, a method used by some builders illustrates how to cut down on framing costs. A wall for R-19 insulation (5½ inches thick) could be framed with 2x6 lumber and 5½ inches of fiberglass placed between the studs. Or, if it figures out to be less expensive, the wall could be framed with 2x4 lumber, with a 1-inch thick layer of polystyrene insulation on the outside in place of the sub-siding. While this may only figure out to R-16, it offers greater protection from cold transmission through the studs.

Building Code Conflict

Some building codes might conflict with this latter insulation design. There should be some type of bracing in every wall in a house. This can be plywood or other rigid sheeting on the exterior, plywood only on the corners, a brace made of dimension lumber either cut into the studs and plates or fitted between them, or a brace made of steel strap nailed onto either the inside or outside or both. A layer of polystyrene on the outside of the wall doesn't give any bracing, so the polystyrene may have to be put on over a layer of plywood. Check the building code closely to see what can be done.

If more insulation is needed than a standard wall will allow, you can use one of the special building techniques listed in chapter twenty-five. For instance, double walls have become quite popular. They can be placed far enough apart to allow any amount of insulation, without the conduction problems a single wall has.

Special planning for insulation is required in the attics of most houses. Because of the design of both trusses and rafters and joist construction, the amount of insulation that can be placed near the eaves is limited to as little as one fourth as much as is needed. If an attic needs R-40 insulation, a strip as wide as 30 inches along the eaves can have inadequate insulation. On a house having a total eave length of 60 feet, this will be about 150 square feet, or the size of a bedroom. Would you underinsulate one of your bedrooms?

Once you have decided how much of what insulation goes where, be sure this is understood by everyone working on the job. From bitter experience I can assure you the various craftsmen involved in building a house are concerned with what someone else does only as far as it interferes with what they are doing, and most feel they will be long gone before any insulation work is done. Everyone who works on the project should be made aware how their job and the others are tied together, especially as far as insulation is concerned.

A prime example of the problems is in the very beginning of the construction, with the concrete work. Concrete contractors don't like to put insulation and/or a vapor barrier under footings and basement floors because it slows the drying time of the concrete, and is hard to keep in place during a pour. This translates to them into extra cost in time, which may cut into their profit. A good way to avoid arguments over this is to make sure all contractors on the job understand they will be responsible for a certain amount of insulation work, and should render their bids accordingly. Your costs will be slightly higher than if they just "did it their way," but keep in mind that insulation pays for itself over time.

5. Work You Can Do

HOMEOWNERS who are building a house or contemplating a remodeling project can save on insulation by putting it in themselves. With the exceptions of blowing insulation into the attic of a house or spraying one of the urethane products or cellulose fiber on the walls, putting insulation into a building is hand work. This is a major reason why building contractors have problems doing some of the insulating as the job progresses. It takes time.

While some parts of the house under construction are more difficult to insulate than others, with the exception of those needing special machines, insulation work isn't beyond the abilities of the average homeowner. If you have a handyman bent, so much the better. Doing it yourself can also result in a better job. You will be living in the house so you will put in more effort than someone who is being paid to do a job.

Equipment is simple and inexpensive. You'll need gloves, cap, and jump suit or coveralls, a dust mask, safety glasses, and perhaps knee pads for working in the crawl space. The tools aren't expensive either. Have a measuring tape, a sharp utility knife (or another type of knife if you prefer) or large shears, a couple of short boards for cross-cutting, and a couple of long boards for lengthwise cutting. If you use a rigid form of insulation such as polystyrene, a chalk line and a drywall square will be very useful. If you are putting up some of the backed types of insulation, a staple gun will also be needed. (If you are putting in a vapor barrier, use only friction fit insulation. This does not have backing. If studs are spaced properly, it will stay in place.)

Unless your construction plans call for rigid insulation, the walls will generally be done with fiberglass, either in the form of batts or rolls. The batts can be either 4 feet or 8 feet in length. Both batts and rolls come in widths made to fit in the generally used spacing between studs and joists, 16 inches and 24 inches.

Cutting Rolls or Batts

To cut batts or rolls, place one board under the insulation at the place you want to cut, and the other one on top. Put your knee on the top board, compressing the insulation, and using the edge of the board for a guide, cut the insulation with the knife. If the insulation has a backing, put that down, away from you. If you use shears, the boards won't be needed. Cut fiberglass insulation 2-3 inches longer and 1-2 inches wider than the opening to insure a tight fit.

Earlier I cautioned against using insulation that is thicker than the space in the wall. When you unroll insulation from a long roll, it will usually be compressed to half or less its listed thickness. Don't worry about this, as fiberglass will expand until it is the thickness it is supposed to be. If you put too thick an insulation in a wall you will not gain the R-value of the thicker insulation, and may even lose some value, and the insulation may expand enough to damage the interior wall sheathing.

Another horror story. One time I was doing a remodel for a homeowner who had purchased a large amount of 3½-inch thick rolled insulation at a bargain. We were furring out basement walls with lumber 1½ inches wide, giving approximately 2 inches of space in the wall. As the insulation was compressed, the homeowner decided we would use the 3½-inch insulation in the 2-inch wall. The insulation went in easily, and we nailed the gypsum board in place without trouble before quitting for the night. The next morning we were chagrined to find the gypsum board badly deformed and in some places pushed completely off the wall. The pressure of the expanding insulation had even pushed it off the nails.

If you have insulation that is too thick for the wall space, it is possible to peel the material into the thickness desired after it has expanded. Unroll it where it can expand, and within twenty-four hours it will be easy to split into the proper thickness.

Cutting Rigid Insulation

To cut rigid fiberglass insulation, place it on a solid base with the backing material up. A sharp utility knife will reach completely through 1-inch material. For the thicker material, a larger knife will work well. Polystyrene and other rigid plastic insulations can be cut by scoring them halfway through with the knife and then snapping them at the cut. Rigid insulations should be cut to fit, so mark them accurately and cut them precisely.

Mineral wool for use in walls is manufactured in batts which are fully enclosed in paper envelopes. These should be cut with shears, as a knife will do more tearing than cutting. Cut them 2-3 inches longer than the space, to allow for settling. Don't cut these lengthwise as the material can get out of the envelope. Use of these for walls isn't recommended.

Leave blowing and spraying insulation to those who do it all the time and have the expensive equipment to do the job. They make it look so easy to get the proper depth and placement that it appears anyone can do it. Not so! This is definitely a place where "practice makes perfect."

For some small blowing jobs, equipment rental companies have small machines a homeowner can use to do good work. However, to do a whole house with one of these takes a long time, for at least two people. A professional insulator can usually blow insulation into an attic in less time than it will take a novice to get ready.

The major thing for a homeowner to remember is to get a reliable contractor for the job. Don't grab the cheapest price, but do some checking around. Ask for names of previous customers, and talk to them. If you have a general contractor doing the whole job, then it's his worry. He should be liable for any problems.

Work You Can Do

There is some work a do-it-yourselfer can do to save on a blown insulation job. Vent/baffles need to be installed between each pair of rafters, at the eaves, to prevent the insulation from blocking the air moving up from the soffits. These are flat sheets of cardboard or polystyrene which are scored to be bent at specific spots to make them fit into the space between the rafters, and also bent down to the top of the wall plate. They are stapled in place, usually 1 inch below the roof sheathing, with the extension stapled to the top of the wall. They may be easily installed before the ceiling is put up, from below. This is a time-consuming job, and you should be able to cut the price of the insulation job by putting them in.

Another job that should be done before any attic insulation is put in is to place the "dams" needed around recessed light fixtures, chimneys, and any place where heat may be transmitted to the insulation and start a fire. The dams are usually made of sheet metal, often aluminum flashing material. They can also be made of scraps of Sheetrock if this is allowed by the local building codes. Some insulators buy the negative sheets from local printers. These are thin sheets of aluminum used by newspapers as the master sheets when printing. Whatever is used, be sure it is approved by your local building code, and that you leave the required amount of space between the dam and the heat source. This is around 3 inches in some codes.

One reason for keeping insulation away from light fixtures is the danger of fire. Some blown insulations will burn, even with fire retardant materials mixed with them. It has also been reported that even if an insulation won't burn, it may transmit enough heat to ignite nearby flammable material, such as wood framing. Some codes allow the complete enclosure of heat-producing items such as recessed lights and bathroom heater/fans, but most just want an open-topped enclosure around them. Check the code for specifics.

Another way to make an insulator happy (which should translate into a cash savings for the homeowner) is to provide an adequate entrance to the attic. Most building codes require some way to get into the attic of a house. The problem is that most builders, and too many building inspectors, believe this must be from inside the house. It is usually a small hole cut in the ceiling of a closet. In an inside closet, "accessible" is open to argument. The main reason cited by the building codes is for fire protection. Trying to get through a closet full of whatever accumulation the homeowner has in it doesn't seem to fit that requirement.

Alternatives

A close reading of the building code will often show that the opening can be anywhere as long as it is big enough for a person to enter the attic. This is another spot where planning pays off. Instead of through the ceiling, why not through a gable, from the outside, or from a garage, if the house has one? In a house with a gable roof, the access may also be an attic vent, taking care of two problems at once. It could also be through a large roof vent, designed to be opened easily. A house with large cupolas in the roof, as in some rustic designs, could provide access through those.

Access through an attached garage is also preferable to going through the ceiling inside the house. A pulldown stair may be installed for quick and easy access. A caution here, though. These stairs are huge air leaks, with no way to seal them. They should be put into a garage ceiling only if the garage is unheated and uninsulated and there is a fire wall (usually only one thickness of gypsum board) between the garage attic and the house attic. Some building codes also require self-closing doors in these fire walls. In the case of attic access, the homeowner should spend some time with the building inspector and get an agreement on what may be done where before a nail is ever driven. Get any information in writing, with names and dates.

Insulating a New Home

Here are step-by-step directions for insulating a new house:

1. Place a vapor barrier and insulation as needed under all forms before concrete is poured.
2. If insulating the outside of poured or concrete block foundations or basement walls, do them as soon as practical after the forms are off. Place sill seal on the foundation wall tops.
3. While framing is in progress, insulate places that may be blocked off, such as corners and where inside walls meet the outside walls. Put sill seal under all outside walls as walls are erected.
4. Once the building is weathertight, install the eave baffle/vents and dams around recessed fixtures. Insulate the small gaps around window and door frames, and seal any large cracks in the framing.
5. Insulate the outside walls. This step may be interchanged with step 4.

6. If you are going to use batt or roll types of insulation in the ceiling, install this. If it isn't easier from the underside, wait until after ceiling gypsum board is in place. (Step 8)
7. Install the vapor barrier, doing the ceiling first, if a vapor barrier is required in your climate.
8. Once the ceiling gypsum board is in place, insulate the attic with blown insulation.
9. If using an all-weather-wood foundation, insulate after all the rough-in work is done in the crawl space. Insulate the rim joists at this time. Install a vapor barrier on the inside of the foundation wall. A vapor barrier should also be installed at this time directly on the ground in the crawl space. If using a concrete foundation, this may also be insulated at this time, if it hasn't been insulated on the outside.
10. In a basement, install insulation and a vapor barrier before the floor is poured. Install insulation in the rim joist spaces before any basement ceiling is put in, to eliminate cold spots. Put in any insulation going on the inside of a poured or block basement wall after the rest of the house has been finished.

6. Installing the Vapor Barrier

A VAPOR BARRIER, also called a "vapor retarder," is a layer of impermeable material usually placed on the heated side of the exterior walls of a house, between the interior wall sheathing and the insulation. Its name states its purpose: to stop the moisture vaporized in the warm air inside a house from traveling through the wall and possibly condensing and freezing on the inside of the outer wall covering.

Unless blocked, moisture inside a house will be carried to the outside when the outside temperature is below that of the inside. If the outside temperature is below the dew point, the moisture may condense on the inside surface of the outside wall covering. If below freezing, it will freeze there in the form of frost, and eventually build up into a layer of ice.

When the outside temperature rises above freezing, this frost or ice will melt and run down the wall, often appearing as water dripping out of the wall. It also is seen as streaks on lap siding. In northern areas where the temperature may stay below freezing for months, ice will build up to a point where it can actually push the siding off the studs.

How It Works

Here is an explanation of the relationship between temperature and humidity that should clarify this. In the air outside a house, as the temperature drops, molecules of moisture in the air move closer together. When the temperature reaches the dew point, at which a vapor begins to condense, the moisture becomes visible as fog. If the temperature drops below freezing, the moisture appears as frost on the ground. Weather scientists say that for all practical purposes there is no moisture in the air at temperatures below 14°F.

Relative humidity is given in terms of percentage by weather reporters. It means the percent of the total amount of moisture the atmosphere can contain at a given temperature. Thus 100 percent relative humidity means the air contains all the moisture it can *at the stated temperature* before the moisture begins to condense into fog. The higher the temperature, the more moisture air can contain before reaching 100 percent relative humidity.

Moisture Needed

Human beings and nearly all other forms of life need moisture in the air for comfort and good health. The proper amount in a house is the subject of much debate between medical professionals and building engineers, among others.

To confuse the homeowner even more, we can be healthy with a lot less moisture in the air than we think we need for comfort. Translated into terms of relative humidity which you can read on your home humidity gauge (hygrometer), this means you can be healthy with a relative humidity as low as 25 percent, but you will be a lot more comfortable with it at 45-60 percent. This can cause problems in a house. The amount of moisture that makes us comfortable can be very detrimental to our house if there is no vapor barrier in the walls and ceiling.

We Provide It

Most of the moisture in a house comes from the occupants. We exhale a large quantity of moisture with every breath. Cooking, bathing, and washing clothes add more moisture. In cold, dry weather we may also operate a humidifier to bring the relative humidity to a level where we feel comfortable. When there is too little moisture in the air, our skin feels dry, and we may get a shock from static electricity if we touch a doorknob or another person.

If there is too much moisture in our house, it may appear as condensation on the windows or as frost where there is an air leak around a door or window or the electrical outlets. Keeping the humidity just below the point where condensation appears on the windows may not feel completely comfortable for us, but it is a lot better for our houses.

Moisture Causes Problems

A horrible example of what can happen was investigated by a building engineer acquaintance. In March he inspected the home of a physician because what appeared to be moisture spots were seen on the ceilings near the outside walls. This is a classic indication of moisture/ventilation problems in the attic. The house had been built the previous summer and was finishing its first winter.

When he examined the attic, the engineer found a band of ice-filled insulation about 30 inches wide along both eaves plus a heavy buildup of ice in the outside walls.

Two construction problems were immediately apparent. The house lacked both a vapor barrier and attic ventilation. What little attic ventilation the builder had put in, in the soffits, was blocked by the insulation having been blown tightly against the underside of the roof when it was installed. With no gable or upper roof vents providing a way out, the moisture that came from the house froze on the bottom of the roof, and where the roof contacted the insulation, in the insulation.

In order to prevent the total destruction of the ceiling, doors were cut into the gables and the ice and insulation were broken into manageable chunks and thrown out before the weather warmed enough for the ice to melt.

The problem was compounded by the homeowner. He believed nothing less that 65 percent relative humidity would be healthy for his family, so had taken steps to keep it that high. In addition to the normal amount of moisture from a family, he had vented the clothes dryer inside the house and was using a large humidifier. By measurement, he was putting over fifty pints of water into the atmosphere inside his house every day.

If a vapor barrier had been put in the house when it was built it would have taken a lot less water to keep the humidity at 65 percent, and a lot less money for repairs.

Look for Discoloration

One of the signs of a moisture problem in an attic is spots of discoloration on the ceiling below. This indicates the lack of vapor barrier in the ceiling as well as inadequate attic ventilation. A properly vented attic will take care of any excess moisture, even if there is no vapor barrier, but too many attics aren't adequately vented.

Another indication of inadequate attic ventilation is peeling and blistering paint on the gables. Moisture in the attic, trying to escape, enters the unpainted interior side of the house siding and pushes the exterior paint off the wood. The same problem on the sides of the house indicates a lack of a moisture barrier in the walls, and a problem with excess moisture in the house itself.

When a house is being built, it is easy to install a vapor barrier. But it's not so easy in an older house. However, there is no excuse for inadequate ventilation in the attic. Ventilation systems can be put in at any time.

Plastic and metal siding have been touted as solutions to avoid siding being damaged by moisture. "Never paint your house again," says one advertisement. This implies the only problem is with the paint itself, not with moisture. No matter what type of siding is used, the moisture problem remains and can only be dealt with by installing a vapor barrier and/or ventilation.

If a house is well insulated but lacks a vapor barrier, the homeowner will not receive the full benefit of the insulation. If moisture, especially in the form of ice, invades the insulation, the value of the insulation is diminished. This may damage the house structure and will increase heating bills.

Not Expensive

Putting in a vapor barrier when a house is built isn't costly. The material itself is inexpensive, and as it comes in large sheets, can be installed with a minimum of labor.

There is only one right kind of material for a vapor barrier. This is some type of polyethylene sheeting, usually called by one manufacturer's trade name of Visqueen. There are variations on the basic material, but they are all similar in function. Some are made in special sizes just for use in standard house construction and are specifically designed for use as a vapor barrier.

Polyethylene sheeting comes in clear, green, or black colors and in thicknesses ranging from 2 mils (.002-inch) to 10 mils. (.010-inch) It is packaged in rolls in widths of 2 to 20 feet. The wider widths are folded before being rolled, so a package containing 2,000 square feet (100 feet long by 20 feet wide) of material will only measure about 6 inches square and 48 inches long.

The best thicknesses for a home vapor barrier are 6 and 10 mils. Lighter materials tear too easily and take more time to put up correctly. Clear polyethylene is sometimes used as a cover for greenhouses, but is good for only one season because it deteriorates in direct sunlight. The colored material is sometimes used for a garden mulch and is made in 24-inch and 48-inch widths especially for that purpose. The colored varieties do not degrade in direct sunlight.

Some types of insulation have a vapor barrier. Roll and batt types of fiberglass are the most used. They have a backing of a layer of kraft paper stuck to the insulation with a thin layer of tar, a very thin aluminum foil, tar, and kraft paper sandwich, or a cloth-like vinyl material glued to the insulation. Kraft paper is the material used to make grocery bags. Aluminum foil is also available for a price. The paper or other backing on fiberglass insulation has folded tabs that can be used to staple it to the studs or joists, and helps seal them from moisture intrusion.

Many Barriers Leak

The main disadvantage to these vapor barriers is that they leak. All, including aluminum foil, have microscopic holes which allow moisture to escape into the walls. All require more labor to install than unbacked (called friction fit) insulation and polyethylene sheeting. Friction fit insulation is less expensive than the backed type.

Installing the Barrier

Putting a vapor barrier in new construction is a straightforward procedure. In colder climates, with a couple of minor exceptions, it is installed on the room (warm) side of the exterior walls and ceilings. It is usually fastened to the studs and ceiling joists with ¼-inch or 5/16-inch wire staples. Professional installers often use electric staple guns, but a do-it-yourselfer can get by with a squeeze-type stapler sold in building supply stores for under $10. Avoid using hammer-type staplers as they tend to cut the material. Since the material must be held tightly in place with one hand while the other wields the stapler, it's easy to injure your fingers.

To install the vapor barrier, clean the room of any obstruction and odd pieces of building material. As the vapor barrier will have to be rolled out on the floor, sweep it clear of any debris, especially nails and sharp pieces of wood.

Cover Ceiling First

Cover the ceiling first. Roll the material out across the room in the direction it will be put up, allowing at least 12 inches extra to lap down the walls at the ends. The material should lap down all walls at least 12 inches to make a tight seal along the upper wall plates when the wall barrier is put up.

Unfold the material carefully, and start stapling it at one side of the room, allowing enough for the lap, and doing one side completely. Work across the room evenly so there will be no extra strain on any single spot. Cover the area completely, including all electrical boxes or other openings. These can be cut out after the interior sheathing is in place and finished.

Walls Are Next

After the ceiling is covered, do the outside walls, starting at the top and working down. Use only enough staples to hold the material in place. Make any joints in the material over a stud or joist, lapping it at least 6 inches. Seal the joints with a silicone-based caulking material, smoothing it out so it will not make a ridge where a joint is. Cover doors, windows, and electrical boxes with the material and leave as much of it as possible in place until the gypsum board or other wall sheathing is in place and finished. This protects the windows, doors, and boxes from getting sprayed with paint or plaster. Lap the corners at least 6 inches and seal them with silicone-based material.

Patch small holes with clear plastic tape such as Scotch "magic" tape. Patch large tears with a piece of polyethylene, cutting it to reach from the studs on both sides of the hole, and gluing it in place with sealer. Tape or sealer may also be used to cover staples and eliminate any small leaks where they puncture the material. A bead of sealer can be put on each stud or joist as the interior sheathing is put up, thus sealing as much as possible the punctures in the vapor barrier made by the nails or screws used to fasten the sheathing.

By covering everything with an unbroken sheet of vapor barrier and leaving it in place until the interior wall sheathing is finished, leaks around windows, doors, and electrical boxes are eliminated. Window and door frames should have had any spaces between them and the studding filled with insulation before the vapor barrier is put up.

Avoid Leaks

Electrical boxes are a major source of air and moisture leaks in house walls. In some houses they leak so much that a cold draft may be felt near them when the wind blows. In extremely cold weather, frost may build up on the wall around them, even with the house at 70°F. inside.

There are several ways to deal with this. One is to have sealed boxes installed. These are expensive, about $3 each, at the time of this writing. Another is to place all electric boxes inside the vapor barrier, and not puncture the vapor barrier with wiring. (This is explained in the chapter on special construction techniques.)

For the owner of a house using standard wiring techniques, a little extra work will reduce the problem to a bare minimum. As soon as the wiring is "roughed in," use a caulking gun with a non-hardening caulking such as silicone (check to see that it is compatible with the insulation on the wiring — that it won't dissolve the insulation), and seal the holes around the wires, both inside and outside the boxes if possible. Also seal any holes through the floor or top plates in the wall framing, or any holes leading outside, such as for outside lights.

Seal Around Boxes

After the interior wall sheathing is in place, seal around the boxes against the wall sheathing material with the same type of caulking. Do this before removing the polyethylene from the boxes. Cut the polyethylene *inside* the box, leaving a narrow strip around the edges. (This won't get in the electrician's way no matter how much grumbling you hear.) Finish sealing the boxes by putting in neoprene gaskets under the face plates. These are sold by building and electrical supply houses.

There are four places where a vapor barrier is not installed on the heated side. These are on the floor of a crawl space, under the concrete floor of a basement, on foundation or basement walls, and in climates where cooling is a greater concern to the homeowner than heating. The reason for placing a vapor barrier on the floor of a crawl space, under a concrete floor, and on the outside of basement and foundation walls is to prevent moisture from entering the building from the surrounding ground. (See chapter three on radon for another reason.)

Moisture entering the crawl space from the ground under it can cause rot in the support members of a house. The same is true of moisture entering through a foundation. Moisture entering through a concrete floor can cause mildew in the rooms above it, damage to carpets or other floor covering, and a cold floor. Moisture entering through a basement wall can also cause mildew, along with damage to any interior wall covering. A vapor barrier in all of these places also helps make the house more energy-efficient.

Crawl Space

A vapor barrier in a crawl space should go from the floor of the house down the foundation wall, across the floor of the crawl space, and up to the floor again. Any insulation put on the foundation with the exception of polystyrene or other rigid types glued to the concrete should be on the crawl space side of the vapor barrier to prevent the insulation from being saturated with moisture from the concrete.

All holes and joints in the vapor barrier should be sealed. The thickest material available should be used, as workmen will be walking on it at different times. Any sharp stones or sticks should be removed, and the ground raked smooth before the plastic is put down. The material can be put in at any time, but it is best to wait until all the work in the crawl space is finished and the building above it closed in to prevent damage to the vapor barrier or water from rain standing on it.

To place a vapor barrier under a concrete floor, the area must first be cleaned, removing any sharp objects. The vapor barrier is laid down, then any steel reinforcing is put in place and the concrete is poured. Care must be taken by the concrete workers to avoid puncturing the vapor barrier. Plastic sheeting under concrete causes the concrete to dry (cure) more slowly, which makes it stronger. (Note chapter seven on insulating under concrete and chapter three on radon gas.)

Coat Wall First

To place plastic sheeting on the outside of a foundation or basement wall, coat the wall with tar or plastic roofing compound, then press the plastic sheeting onto that while it is still sticky. The black polyethylene in 10 mil thickness is best for this. All joints should be lapped and sealed. (See chapter seven on insulating these areas for more information.)

Be sure to patch any holes in the vapor barrier, and lap and seal all joints. The idea is to close off all avenues for moisture to enter your house.

In climatic areas where cooling is more of a problem than heating, serious questions rise about installing a vapor barrier on the inside of a house. In some humid areas, it is thought this will cause moisture from the outside air to condense in the walls and cause damage to framing members of the house, especially if the house is air-conditioned. For these areas, engineers have experimented with installing the vapor barrier on the outside of the walls, thus preventing moisture in the air outside from getting into the wall. This also makes the air conditioner work more efficiently, and lessens the load on the dehumidifier. In some areas of the southeastern United States, a few building codes require a vapor barrier on the outside, especially if the building is cooled mechanically.

Need A Vapor Barrier?

Building engineers have established a simple standard that a homeowner may use if in doubt about putting the vapor barrier on the outside or installing one at all. If the *average temperature* for January is 35°F or higher and the cooling degree days greatly exceed the heating degree days, a vapor barrier isn't needed, or should be put on the outside if the area is very humid.

The map on the facing page shows the line below which the barrier isn't needed.

A homeowner living on the south side and close to this line should check the local weather service office to find out if the 35° rule applies. It is better to have a vapor barrier if there is a question.

Moisture Line

Engineers have drawn another line to guide those concerned that moisture in the walls of their houses will cause damage. Homes east of a line through Texas, from the southernmost tip, north and gradually swinging east to the northeastern tip of Minnesota are in a "moderate decay prone area." The extreme southeastern area of the country is listed as a "high decay-prone area." If you live in those areas, get maximum wall and attic ventilation, even if you don't use a vapor barrier.

Another product on the market has been mistaken by some homeowners for a vapor barrier. The usual name for it among builders is "house wrap." It is a tough, paper-like material that is put on the *outside* of a house, under the finish siding. It is designed to prevent air intrusion, better described as the wind blowing through the walls. It may be put on over the sub-siding, or directly onto the studs if there is only one layer of exterior siding.

Unlike a vapor barrier, which must be moisture-proof to function correctly, house wrap must allow moisture-laden air to pass through it. The special material now on the market does this, *one way!* It allows any moisture in the walls to travel to the outside, but doesn't allow moisture to travel back into the walls. It must be put on right side out, otherwise it will allow moisture into the walls from the outside.

Over the years different materials have been used to wrap the outside of houses and try to prevent the wind from blowing through the walls. The most common were a heavy paper called "building paper" and a similar material soaked with tar, called "felt paper." Both allowed moisture to travel in either direction. They did help keep the wind from blowing through, but didn't keep moisture out of the walls completely.

Some older houses have aluminum foil under the exterior siding. This was often placed as a radiant barrier (see the chapter on insulating for

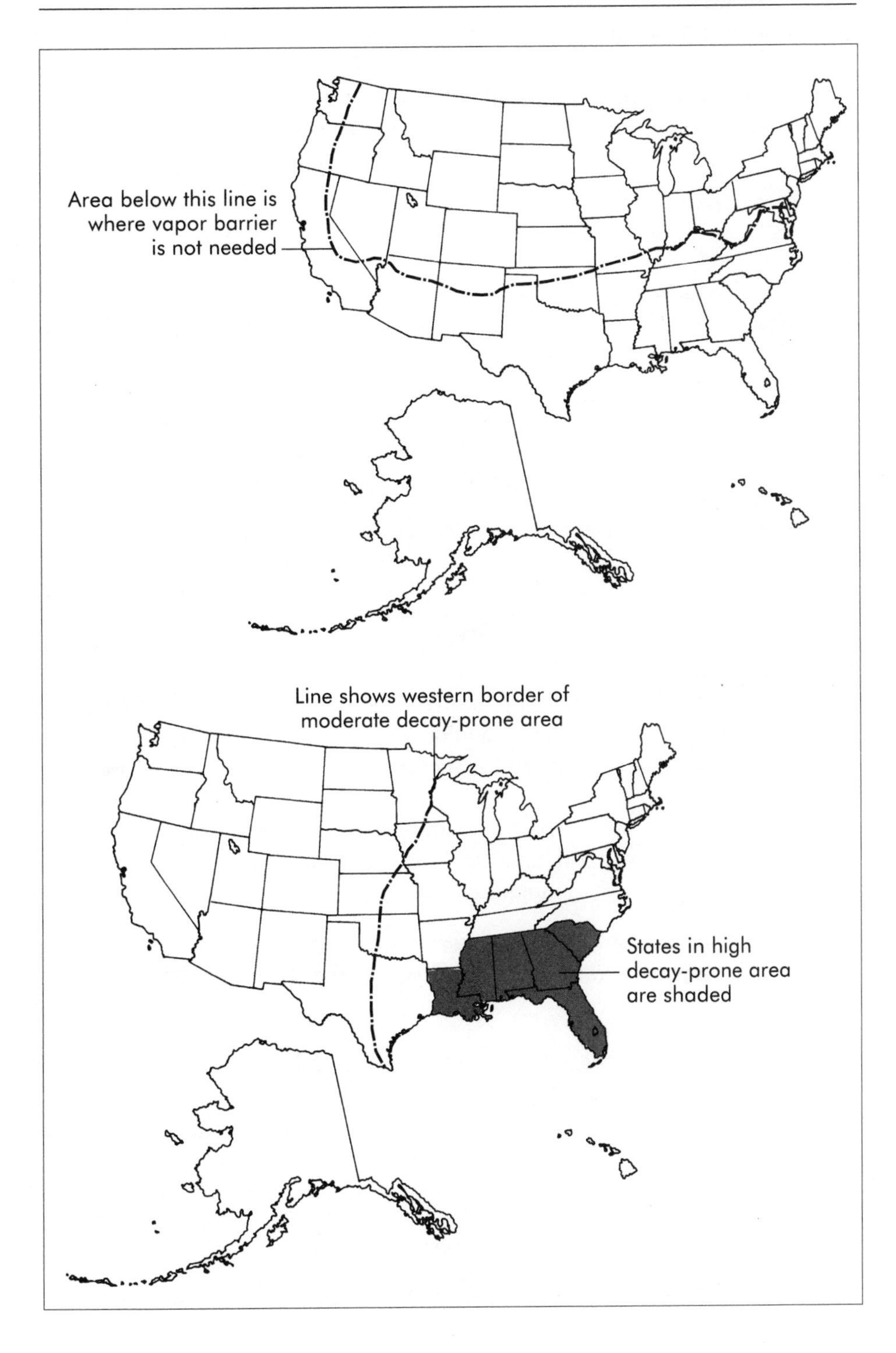
Area below this line is where vapor barrier is not needed
Line shows western border of moderate decay-prone area
States in high decay-prone area are shaded

Humidity Chart

Chart of increase of weight of water by grains per pound of dry air as temperature increases at 20 percent, 40 percent, and 60 percent relative humidity. (7000 grains = 1 pound)

20 percent relative humidity

8°F	20°F	40°F	60°F	80°F	100°F
0	4 grains	7	16	30	59

40 percent relative humidity

0°F	20°F	40°F	60°F	80°F	100°F
0	6 grains	15	31	62	118

60 percent relative humidity

-20°F	+20°F	40°F	60°F	80°F	100°F
0	9 grains	21	46	90	169

(Information derived from psychometric chart published by Wing Company. All figures for sea level.)

cooling) to keep the sun from heating the inside of the house too much. Because it is to some extent impermeable, it also keeps in the walls moisture from the inside of the house. During cold weather this will freeze and cause problems.

House wrap, if installed properly, will help a homeowner save energy dollars. As with insulation and the vapor barrier, it is part of a much larger picture, a house planned for energy efficiency before a nail is ever driven.

The table below shows how a family of four can produce over 24 pints of water a day. If this much water were poured out on the kitchen floor it would take a lot of mopping to clean it up. Yet this much and more

Moisture Produced by a Family of Four Each Day

Activity	Moisture produced (In Pints)
Cooking three meals per day	1.9
Dishwashing three meals per day	.95 (each time)
Bathing (shower)	.95 (each time)
Bathing (tub)	.095 (each time)
Washing clothes (per week)	3.8
Drying clothes with unvented dryer (per week)	24.9
Occupants (family of four, per day)	11.5

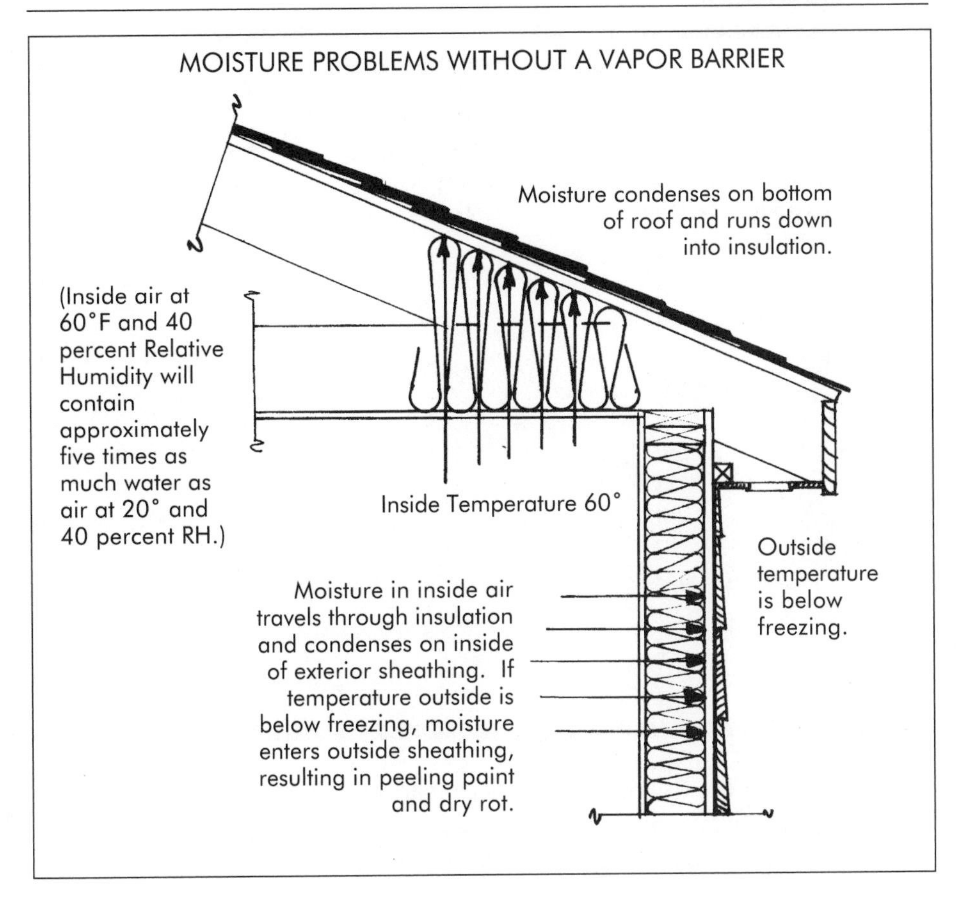

can be added to a household atmosphere almost daily without the occupants being aware of it, simply because they can't see it.

In addition to the listed activities, pets, plants, aquariums, and self-defrosting freezers and refrigerators all add moisture to a house.

Setting the Humidistat

The proper relative humidity for your home depends upon factors such as outdoor air temperature, effectiveness of weather stripping, type of windows and doors (including frames and jambs), and whether storm windows and doors are used. With all these variables it is nearly impossible to recommend a proper humidity setting. The best humidistat setting is one that you are comfortable with. Also, as the outdoor temperature fluctuates, it may be necessary to adjust the humidity level of your system a few times during the heating season.

Relative Humidity Chart

Outside Temperature	Outside Relative Humidity	Inside Humidity (%) When Outside Air is Heated to 72 Degrees F	Maximum Safe Recommended Relative Humidity
-10 Deg. F	40	1	20
	60	2	
	80	2	
0 Deg. F	40	2	25
	60	2	
	80	5	
10 Deg. F	40	4	30
	60	5	
	80	7	
20 Deg. F	40	6	35
	60	8	
	80	11	
30 Deg. F	40	8	35
	60	13	
	80	17	

Refer to the Relative Humidity Chart as a starting point for your proper humidistat setting. Generally, in a tighter and better insulated house, the humidistat may be set higher than in a drafty, uninsulated house. *Do not overhumidify.*

Because cold air cannot hold as much moisture as warm air, any cold drafts or cold windows and doors, including the frames, may cause water vapor to condense at these points. Also, if your home is well-insulated and weather stripped but lacks an effective vapor barrier, water may seep through the walls and ceilings. This moisture may condense either inside or on the outside of walls or in the attic. If any of these conditions are observed, the humidity should be reduced before water damage occurs.
(Chart and information courtesy of American Metal Products Co.)

Insulating New Homes

7. Foundations, Basements, and Crawl Spaces

IF YOU LOOK AT HOUSES you will see that the basement or foundation walls of most of them aren't insulated on the outside. By being the exception to the rule, you can greatly increase the comfort of your new home and reduce energy bills to a minimum.

The main reason this isn't normally done is the cost factor, mostly to the contractor doing the work. These "extra details" take extra time. To a contractor, time is money. Extra time means extra cost, which might be from his profit. Better not to spend extra time, he thinks.

You will notice in the chart on R-values in chapter two that concrete has little resistance to heat loss. In fact, it is an excellent conductor of cold. By placing insulation between the source of the cold, such as the surrounding soil, and the house, you prevent the basement and foundation walls from conducting cold into your dwelling. Insulating basement floors and walls or foundation walls against the cold will pay off in greatly increased comfort within, and no small consideration, in reduced energy costs to the owner.

Frost May Show

A concrete or concrete block basement wall may get cold enough to show frost on the living side even with the temperature in the basement in the 65°-70°F range. I have worked in basements where a band of frost showed on the inside wall at the depth of the frost in the ground outside. The room temperature, as shown by a thermometer in the center of the basement, was 70°F, but the thin layer of ice on the concrete block basement walls never melted.

Concrete in direct contact with cold soil or air will be virtually the same temperature as the cold soil, despite efforts to heat the air on one side of the wall. It has been claimed that in many temperate areas, after a certain depth, soil temperature remains a constant 55°F. This is one of the reasons for constructing an earth house, where the building is partially or totally buried. Since comfortable living temperature is figured at 65°-70°F, the reasoning is that it will only require enough heat to warm the room 15°-20° instead of the possible 70°-100° needed in a standard above-ground house. Therefore, there will not be such a large expenditure for heating or cooling.

The reasoning is sound, but some other factors enter into the equation. If the soil contacting the concrete floor of a basement is 55°F, the concrete will be at or very near the same temperature and 55°F feels cold when your feet are at that temperature all the time.

Needs Insulation

For those reasons, insulation and a vapor barrier should be placed on the soil side of all concrete basement walls and floors. Insulated from cold, concrete will absorb heat, contributing to the comfort of the house instead of detracting from it. (See chapter eighteen on insulating earth houses.)

It's easy to insulate basement and foundation walls on the outside during construction. The excavation for the basement or foundation must be larger than the outside perimeter of the walls because room is needed to place the forms if the walls are poured concrete, or for the blocklayers to work if they are block.

Part of your planning should have been to determine how much insulation would be needed (R-value) for the basement or foundation walls, so you must decide only what insulation to use to get that figure. The quickest way is to have the walls foamed with one of the types of urethane available. This may not be the least costly. After urethane, the best material to use is polystyrene. Both materials are waterproof, but should still be protected from the soil by a layer of polyethylene.

Before any concrete is poured, at least a layer of polyethylene should be put down. In a climate requiring insulation of basement or foundation walls, an equal amount of insulation should be under *all* concrete.

In very cold climates where the frost may penetrate below the depth of a foundation or even a basement, the excavation may be deepened by

12-24 inches. Then a layer of polystyrene should be placed first, topped by a layer of dry crushed gravel laid up to the depth of the bottom of the footings. Another layer of polystyrene is put on the gravel, then the concrete is poured. This can prevent any shifting of the basement or foundation from action of frost, and helps to make a crawl space much warmer.

Polyethylene sheeting placed under any footing should be wide enough so it can be folded around and brought up the wall above the footing at least two feet. This is to allow the polyethylene sheeting on the wall or in a crawl space to overlap and seal against any intrusion of moisture. (And cold!)

Once the forms are stripped from fresh concrete, construction of the building can begin in from three to five days. This gives the concrete time to harden so it can't be damaged easily. Also, most builders won't backfill around "green" concrete walls until there is enough weight from the construction of the building to keep them from bending and cracking from the weight of the soil around them or the equipment used for the backfill job.

Insulating on Outside

It is easy to insulate the basement or foundation walls on the outside prior to backfilling around them. The first step is to coat the wall with a layer of black plastic roofing compound or similar material. This will seal any small openings that might leak water from the outside. It is also required by many building and lending codes. While this coating is sticky, press a layer of polystyrene boards into it. If more than one layer of polystyrene is required, glue the second layer to the first with one of the caulking gun glues made for that purpose.

When the polystyrene is in place, cover it with black polyethylene sheeting, either letting it hang until held in place by the backfill material, or gluing it with the same type of glue used on the insulation. Cover any punctures with a patch of the same material held in place with glue.

Protect the Insulation

Also needed if a basement or foundation wall is insulated on the outside is some way to protect the insulation in the area between the ground level (finished grade) and the house siding. There are several products just for that purpose. One is a vinyl-covered polystyrene. The vinyl covering makes a good-looking finish for a foundation, and protects against damage to the insulation. There are also some types of "cement board" which will protect the insulation, but still look like concrete. Another product is a polystyrene insulation surfaced with a hard finish that looks like stucco. It

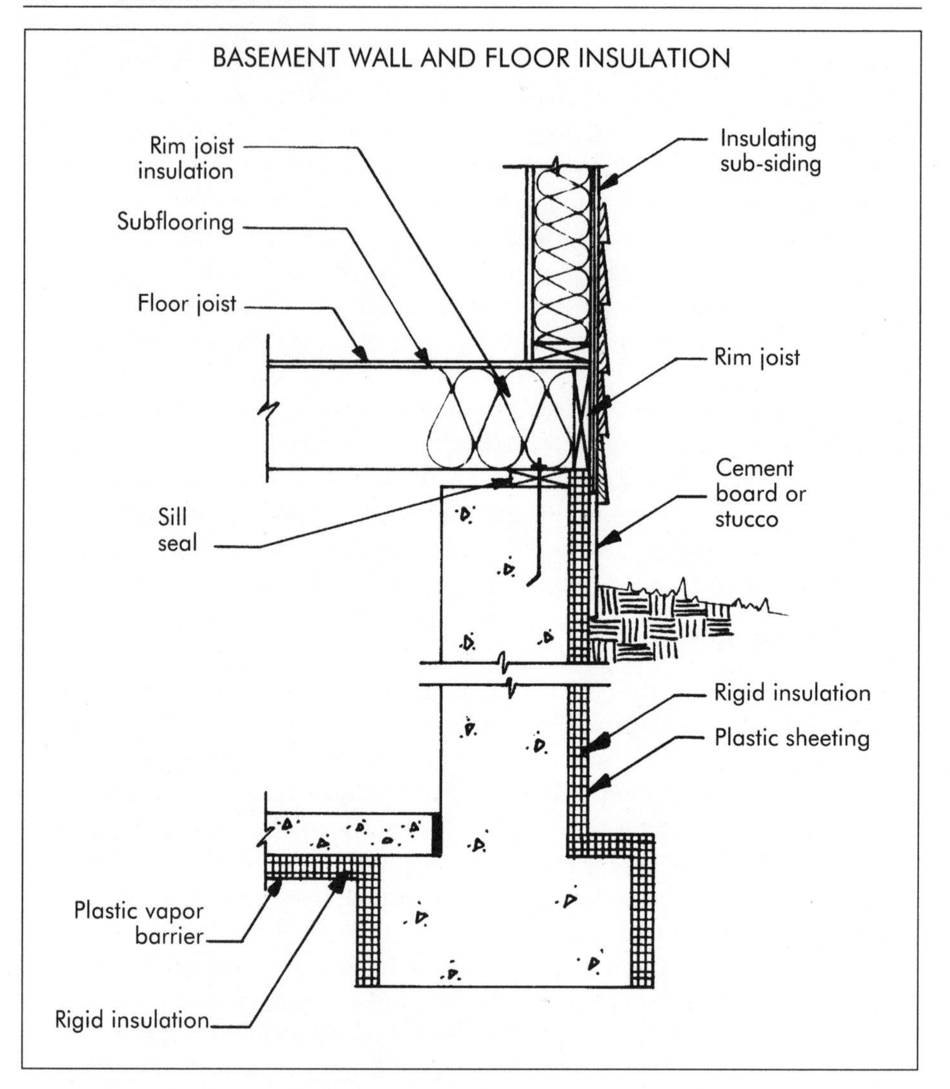

may take some shopping around to find what you want, but that's all part of having a new home.

In areas with termite problems, and they extend farther north than most people think, a termite shield may be required by the local building code. This is usually a thin sheet of copper, copper-coated steel, or galvanized steel placed on top of the foundation or basement wall. It is wide enough so there will be an overhang on both sides of the wall of 2-3 inches. Check your code for specific dimensions. This should extend beyond any insulation on either the inside or outside (or both) of the wall.

A bead of caulking such as latex/silicone or similar non-hardening material should be put under the termite shield to prevent any small air leaks. Sill seal can also be used if allowed by the building code. In any case, sill seal should be placed on top of the termite shield and under the mud sill, the board (usually of all-weather wood) that is bolted to the top of the foundation and serves as the base for all construction above it.

There have been reports of termites tunneling through polystyrene insulation when it is damp from soil moisture. This is another reason for the polyethylene sheeting on the outside of any outside insulation.

In your planning, allow for the thickness of the insulation on the outside of the foundation walls so your house walls will cover the top of the insulation. In other words, the circumference of the foundation walls *with insulation,* should be slightly smaller than the completed outside dimensions of the finished house so no water will run down the upper wall and into the insulation. Ideally the finish siding on the house should extend slightly below the top of the insulation and any finish you put on it, to make a weathertight joint.

Crawl Space

When there is a crawl space under a house, building codes usually require several vents or screened openings through either the foundation or the rim joist (the board that rests on edge on the mud sill), which has the floor joists nailed onto it. These vents allow air to circulate under the house, removing any moisture that might accumulate and cause damage to the underpinnings of the house.

There is an obvious conflict here. If the foundation wall is insulated to prevent cold from penetrating under the house, then vents are left open to air circulation, it appears the insulation will do little good. That is correct. If the vents are left open, it is unnecessary to insulate the foundation. But you must insulate the floor and all water and sewer pipes in the crawl space, along with all heating ducts.

The alternative to that expensive problem is to insulate the foundation wall, close the vents during the cold months, place a good vapor barrier on the floor of the crawl space to prevent moisture intrusion from there, and maybe put some source of heat in to keep the floor warm.

Improved Vents

The vent manufacturers have foreseen this problem and are making crawl space vents with an easily closable shutter. All the homeowner has to do is walk around the house twice a year, closing the vents in the fall and opening them in the spring. Otherwise the homeowner will either have to enter the crawl space and put some insulation over the vents, or cover them from outside. There's one exception to closing *all* vents for the winter. An opening is required for combustion air for an oil- or natural gas-fired furnace or water heater in the crawl space.

Estimating Vent Area

In the absence of a specific local building code requirement, the rule is to allow 1 square inch of vent area for each 4,000 Btu's rating of the appliances. (For example: The furnace and water heater add up to 100,000 Btu's. Twenty-five square inches of vent area will be needed.)

Some codes require this be halved, with half entering the "room" within 12 inches of the ceiling and the other half entering within 12 inches of the floor.

In a crawl space, this can be achieved by putting a round warm air pipe elbow over half of the vent, with a piece of pipe extending to within 12 inches of the floor of the crawl space. The other half of the vent is left open directly into the crawl space. This opening should be well away from water or sewer lines to prevent freezing.

This problem can be circumvented in two ways, by installing the new closed combustion appliances, which have a sealed pipe leading directly from outside the house, or by putting the appliances in some part of the house where they can get outside air without puncturing the tight envelope of the house, in a room with an outside entrance, for instance.

Insulate Rim Joist

Once the building is closed in and weathertight but before any interior finishing is done, the rim joist should be insulated. This is the space between the floor joists from the sill on top of the foundation or basement wall to the subfloor. The insulation should extend across the width of the foundation wall. It may be insulated with polyurethane foam, as is recommended for the "best" job, blocks of polystyrene, or stuffed with fiberglass. A reason for using foamed polyurethane is to stop air leaks along the rim joist. This may be done, if necessary, by caulking the top and bottom of the rim joist from the inside, or running a bead of caulking material on both the top and bottom edges of the rim joist as it is put in place.

If for any reason you cannot insulate the outside of a basement/foundation wall, turn to the chapter on insulating the basement/crawl space on older homes for information on how to insulate from the inside.

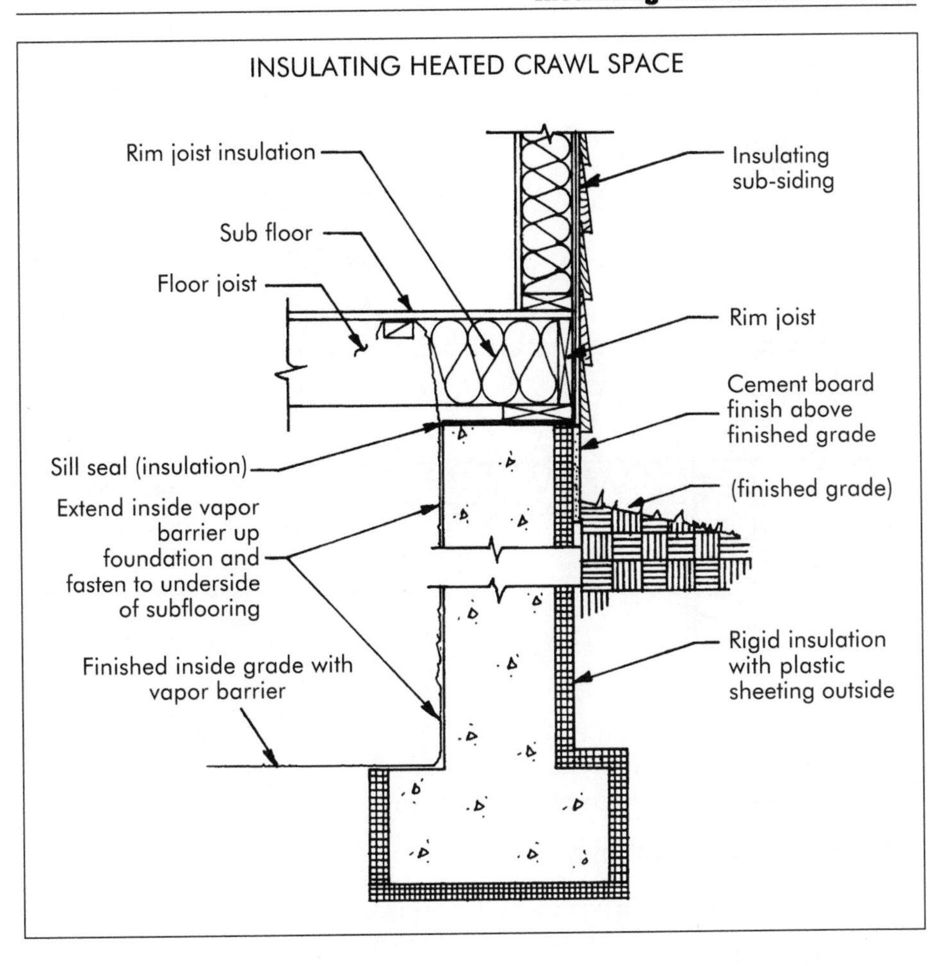

8. Floors

SOME BUILDING DESIGNS call for insulating part or all of the floor. This is a special problem, and isn't solved by simply stuffing insulation between the floor joists.

Houses constructed with cantilevers (overhangs) have part of the floor of the living area exposed to the outside weather. If some thought isn't given to the placement of insulation in those areas, there will be cold areas in the floor and a source of cold that the heating system must overcome.

Even if the floor joists are deep enough to hold the required amount

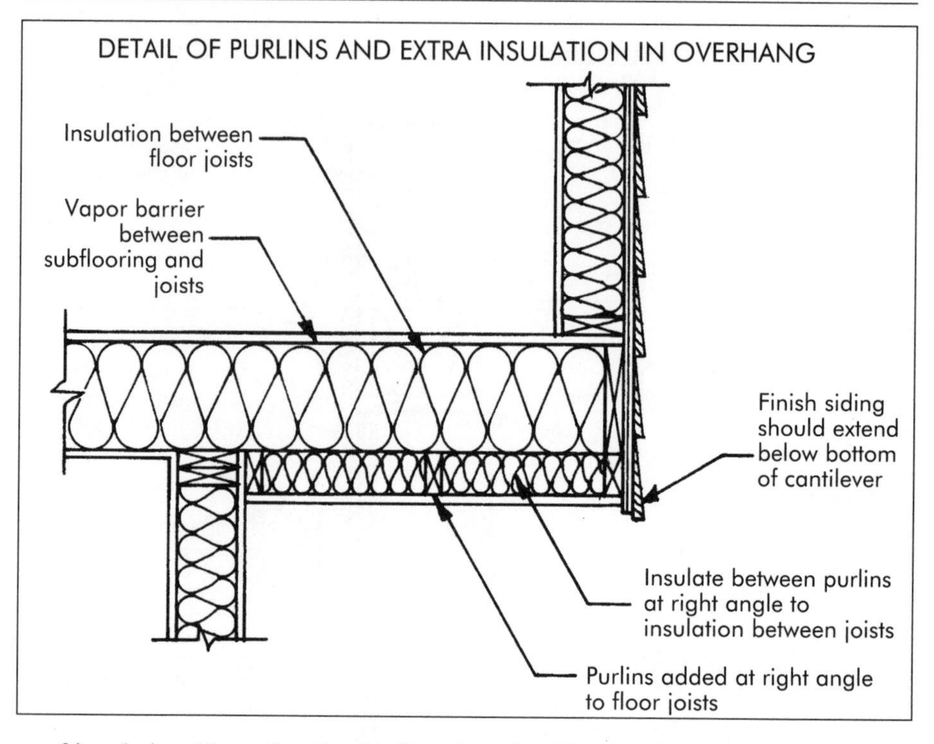

of insulation (three-fourths the R-value of ceiling insulation), there is still the problem of cold being transmitted through the joists.

Two Possible Solutions

This can be overcome in two ways. One is to build another layer of insulation onto the bottom of the joists, at right angles to them, so the area of cold transmission through the solid wood is reduced to an absolute minumum. With two layers of insulation also at right angles to each other, any small cracks are also reduced to a minimum.

An easier method is to fasten a layer of rigid insulation onto the bottom of the joists, and then cover that with plywood or flake board held in place by long screws that reach into the joists. With this layer of rigid insulation at least 2 inches thick, transmission of cold through the joists will be reduced to a minimum.

Any solution to this problem, either the use of purlins fastened to the floor joists or rigid insulation between them and the outside sheathing, must be part of the plans for the building. If heating ducts are located in the cantilever, extra insulation will have to be placed to keep them warm. If they are not insulated properly, there will be a short blast of cold air every time the furnace blower starts, and more cold for the heating system to overcome.

Another item usually overlooked when a floor is insulated is the vapor barrier. As with the walls and ceiling, this is placed on the warm side of a floor. That means it will have to be put on top of the joists before any subflooring is laid, or placed between the subflooring and the under-layment. Either way, it will be subject to damage while the house is under construction.

The underside of the cantilever should be covered with ¾-inch plywood to prevent damage to the insulation in the floor.

Floor Over Crawl Space

Floor insulation is also used in some areas of the United States where moisture in the crawl space is a problem. The idea is to insulate the floor, and allow air to circulate through the crawl space under it. This is another place where builders usually just stuff the insulation between the joists and let it go at that.

Several problems must be taken care of if the house is to have a warm floor. The first is the possibility that the floor joists won't be high enough for the required depth of insulation. Then there is the vapor barrier. It must be put on the warm (room) side of the insulation. As with cantilevers, there is a possibility it will be punctured when the subfloor is being laid. When the underlayment is put down, the fasteners (usually staples) must be put into the joists so they won't puncture the vapor barrier even more.

Once the insulation is between the joists, some builders use only wire mesh to hold it in place. Then if the crawl space is left open all year round, just the wind may damage the insulation. Mice will also penetrate it and, in some cases, haul it away to build nests. If a vent screen is damaged or removed, birds will also haul insulation away for nesting material.

If a floor is insulated in this manner, the underside of the insulation should be covered with some type of rigid material such as plywood or particle board to protect the insulation from damage. The floor of the crawl space should be covered with polyethylene sheeting to prevent moisture from rising up into the insulation from the ground under the house.

Protect Plumbing

In a normal single-story house, most of the plumbing is in the crawl space. This will have to be insulated or located on the warm (room) side of the insulation to prevent it from freezing. Even if you can convince your plumber to put the pipes near the floor between the joists, it will be expensive. Other problem spots are the water supply line entering the house and the sewer line leaving it, both passing through the cold crawl space. We will discuss insulating these in a later chapter.

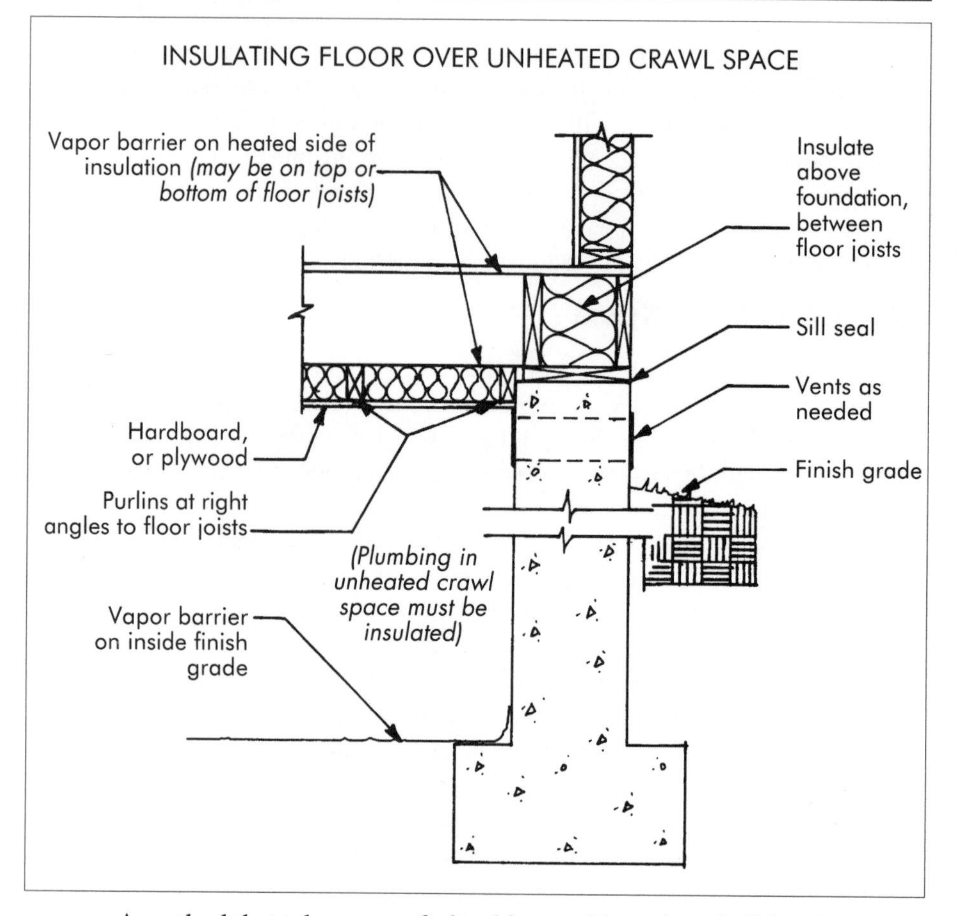

A method that takes care of plumbing problems is to build an insulation floor several inches below the floor of the house, so that most of the plumbing is between that insulation and the floor joists of the main floor, right where most plumbers put it. Put the vapor barrier on top of the insulation floor, use as much insulation as necessary, then put solid sheathing on the bottom to protect the insulation from damage. Crawl space vents will have to be below this insulation floor.

This is the same system used in the far north, where houses are often built on pilings. In that case, the insulation serves two purposes. It keeps the cold out of the house and keeps the warmth from melting the permafrost under the house and causing it to sink.

Houses built on pilings in the southern part of the United States can benefit from this type floor insulation too, but for cooling rather than heating. While no vapor barrier is necessary, plumbing and air conditioning ducts must be protected from the heat. As with heat ducts, cooling ducts must be kept as close to a constant temperature as possible, to reduce energy costs.

If the design of a house or other circumstances require that the floor be insulated, more effort must be put into the job than in any other single part of the house. If this effort isn't made, the result will be a floor that leaks heat. Further problems will be experienced with the plumbing unless special steps are taken.

Briefly, if there is any way to avoid having to insulate the floor of a house, use it in preference to doing the floor. Insulating a crawl space may seem more difficult than insulating a floor, but a comparison of the costs will show it is less expensive and those doing the job will tell you it is less work.

9. Exterior Walls

ONCE THE BUILDING is closed in, and any wiring or plumbing in the exterior walls has been roughed in, it is time to insulate the exterior walls. Your planning and step-by-step insulating should have taken care of the hard-to-reach spots by doing them as construction progressed. By this time, all you should have to do is put insulation between the studs in the outer walls.

Caulk the holes in the electrical boxes before the insulation is put in the walls. It may be done after, but it's more difficult. In any case, do it before the vapor barrier is put up. That way you won't be fighting tightly stretched plastic sheeting when you try to get into the small space in the boxes.

You will notice that there are a lot of spaces between the studs where the insulation won't fit without being cut.

When cutting batt or roll insulation, allow an inch or 2 extra width and 2 or 3 inches extra length. The insulation will fit tighter in the space, and eliminate voids (small cracks) along the sides of the insulation. If you are using rigid insulation, be precise in cutting, as it cannot be compressed and may be damaged if forced into too small a space. Some insulation applicators put a bead of caulking material along the edges and ends of rigid insulation to seal against the wood framing, eliminating any voids.

Foam Insulation

Another method of cutting air leaks in an outside wall is to have a thin (less than 1 inch thick) layer of polyurethane foamed into the wall first, then finish the job with fiberglass. This cuts down on leaks, but is expensive. If a house wrap was used on the exterior, under the finish siding, it probably won't be necessary. Because of the cost of foamed insulation, it should be used only in hard-to-reach spots, or where there isn't space enough to put in the necessary thickness of fiberglass.

Again, a warning about forcing fiberglass into too small a space. If you have a 12-inch space between studs, don't try forcing a 16-inch width of fiberglass into it. Mashing fiberglass insulation reduces its R-value. As it expands over time, it may also force the interior sheathing off the studs, leaving nail holes in the material. Also, don't try to put 6-inch insulation in a 4-inch wall for the reasons just given. Use the thickness of insulation made for your wall. If you want more R-value in a wall, use an insulation with a higher R-value per inch, or build the wall thicker. Plan ahead!

Insulate Around Wiring

You will notice when you start insulating a wall that the electrical wiring gets in the way. Wiring is usually strung through holes drilled midway through the studs and 12 to 16 inches above the floor. The outlet boxes are nailed to the studs, at about the same height, with the switch boxes 48 inches above the floor.

If the building is wired that way, it will take a little extra work to get a good insulation job. To get around the wires, slice into the insulation at the height of the wires, cutting it about half way through. Carefully slip the insulation around the wire so it isn't compressed by the wire, but surrounds the wire and fills the wall space behind it. The insulation will expand and close the cut, not leaving places with less insulation than the rest of the wall.

The usual method of insulating around outlet and switch boxes is to force the insulation behind them, compressing it somewhat. A better way is to use shears. Make a shallow cut, the size of the switch box, that will eliminate compressing and not leave spaces around the boxes with a lower insulation value than the rest of the wall. Take extra pains in putting insulation behind electrical boxes. This is where most large voids occur. Also, if you don't seal the boxes, or use the newer ones that seal the wire, voids behind the electrical boxes will allow cold to come in around them.

Plan Wiring

You can avoid the problem with wiring by a little planning. Instead of having the wires traversing the wall from side to side, through holes drilled in the studs, have them brought into the wall from either the top or the bottom, through holes drilled next to the stud the box is mounted on.

This allows the insulation to enclose the wires without any cutting and makes the insulation job go faster. (The chapter on alternate methods of construction will show you how you can avoid any problems with either the wiring or the boxes.)

Either the first or last step in insulating the outside walls of a house will be to fill the gaps around window and door frames with insulation. Use scraps from the rest of the job, pushing them into the gaps with a thin strip of wood. Better yet, use one of the aerosol products made for that purpose. Whichever way you choose, don't skimp. Fill these spaces completely. It will pay off in comfort for you for years to come.

10. Ceilings and Attics

IN MOST SINGLE-STORY and two-story houses the attic is not heated and the insulation is placed on the ceiling, directly above the living space. This is a saving in several ways. It uses less insulation than would be required to do the roof. The insulation will most often be blown in. This is less expensive than the alternatives. The heating area is also confined to the living space, with no heat being wasted in the unused attic area.

Here is where planning can pay off. Much of the preliminary work for insulating an attic may be done from below, before the ceiling is in place. This takes less time and effort, which translates into a saving for the homeowner.

If the attic is low, as with a 2-12 or flatter pitch (a slope on the roof of 2 inches rise in 12 inches of run), the insulation must be put in from below, and then the vapor barrier and the ceiling put up. Sometimes it may be cheaper to have a steeper roof, where the insulating can be done from above, than to have a shallow attic requiring more work and more expensive insulation. This should be worked out before a nail is driven.

All ventilation in an attic should be installed prior to insulating. It is easier to do the work as the house framing is done, and it is necessary for the safety and comfort of the insulators. Even on a cool day an unvented attic can be hot. Professional insulators usually wear coveralls that are warm, but needed for protection against the insulation and other physical hazards. An overwarm attic can slow their work and may be a cause of a sloppy job.

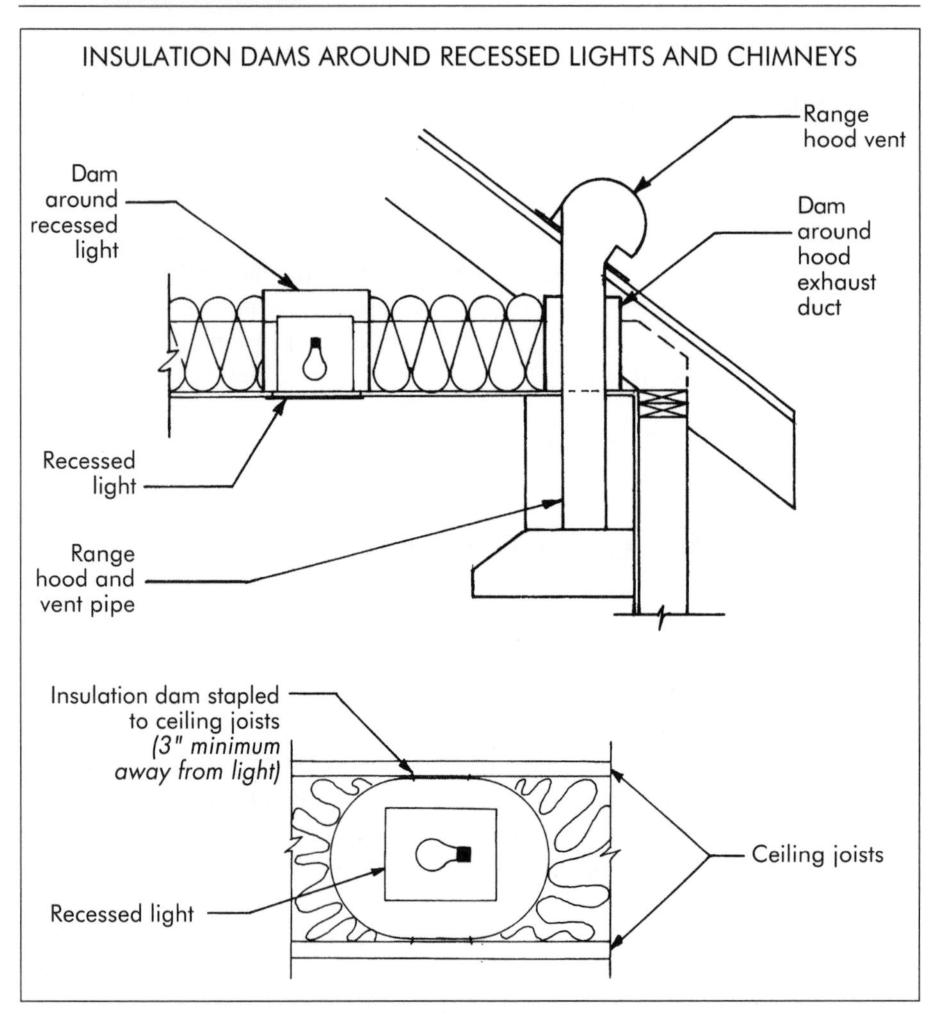

Install Vents

Install insulation baffle/vents before insulating an attic. It is easier and less costly to put them in from below, before the ceiling is in place.

Another preliminary job is to install dams around recessed light fixtures, chimneys, and the range hood vent pipe. This is also easier if done by working from below, especially near the eaves.

Something to watch for here. Some builders may vent the bathrooms and the kitchen range hood into the attic, not up through the roof. Venting bathroom exhaust fans into the attic puts an excessive amount of moisture into the attic space. This can damage the insulation as well as the wood framing members and trusses where the insulation is cellulose fiber made with aluminum sulfite powder.

Venting a range hood exhaust into the attic adds moisture and may cause a buildup of grease from cooking. This is a fire hazard. A range hood should be vented outside, through 26 gauge or heavier stove pipe, which should not touch any flammable material. While most insulations aren't flammable, they may transmit heat to something that is.

Roof or ridge vents should have been installed as the roofing was put on. It is much easier at that time. If the roof vents weren't put in when the roofing was put on, they should be installed before insulating, for the reasons given earlier. Soffit vents should be put in when the soffits are. These do not have to be done prior to insulating if you are using baffle/vents or one of the construction techniques shown in the accompanying diagrams.

If the eaves are to be left unfinished, with the rafters exposed, bird screens at the top of the outside walls between the rafters should be installed from above as the roof sheathing is put on. Bird screen is the wire mesh put between the rafters to prevent birds and bats from entering the attic. Eave venting should be checked after the insulation job is done to make sure it was not blocked by insulation.

The local building code or your lender may require a minimum square footage for the attic vents. The amount of vent area will vary depending on how steep your roof is. Usually the steeper the roof, the less vent area is required.

Formula for Vent Area

Here is a formula that will provide a minimum vent area if the building code or your lender doesn't have one:

> If the roof or ridge vents are at least 3 feet higher than the eave (soffit) vents, the total *minimum net free area* of *all* vents should be at least 1 square foot for each 300 square feet of *ceiling*.
>
> If the roof vents are *less* than 3 feet above the eave vents, the *minimum net free area* of ventilation is 1 square foot for each 150 square feet of *ceiling*.

Net free area is not the same as the area of the opening in the vent. The net free area will be stamped on the vent or be given in the manufacturer's literature. (Ridge vents, a continuous vent placed at the peak of a roof, and soffit "strip vents" are often sold this way.) While it is easy to compute the opening in a baffle/vent or one left where bird screen will be used, any screening cuts that figure down, so find out what the

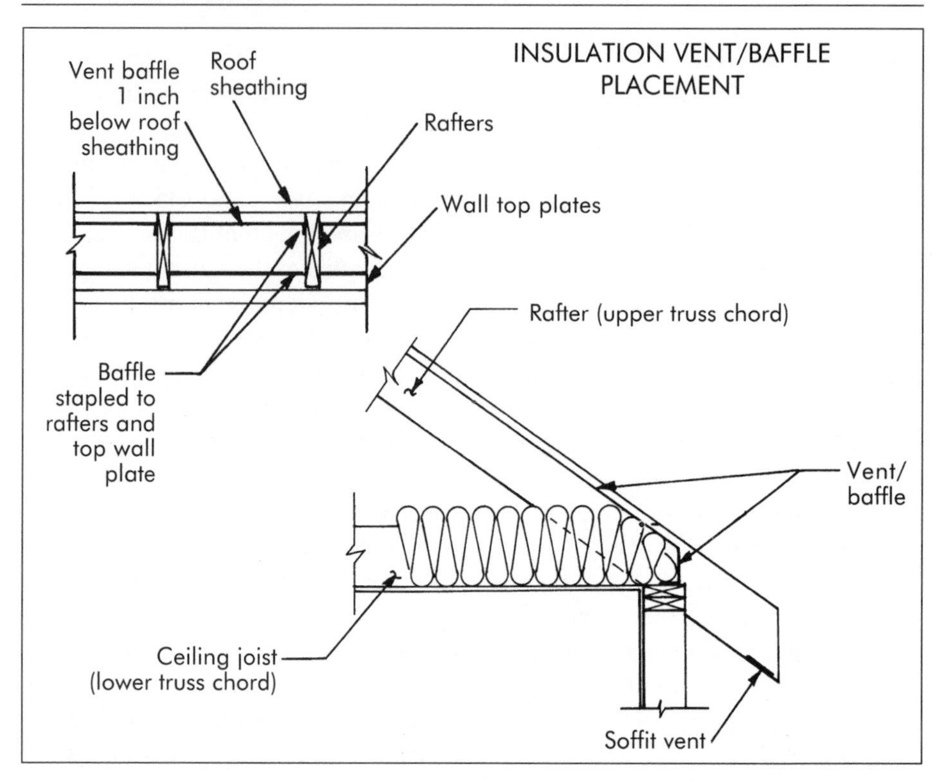

manufacturer of that material says it is. Building inspectors can get finicky about such things. It is simpler, and less costly, to plan to have much more vent area than the minimum you can get by with.

Vent Sizes

The upper and lower vents should be nearly equal in size, or the upper ones should be larger, to allow a good flow of air between them. If you need 6 square feet of net free area of venting, then you should have 3 square feet at the eaves and 3 square feet at the ridge. The size of vents is usually given in square inches, so you will have to compute the square footage from that.

In a ceiling where the insulation must be installed from below, a solution to another problem must be thought out beforehand. Many builders use one layer of roll or batt type insulation, placed between the ceiling joists, to do the job. In a climate requiring only an R-11 insulation value in the ceiling, this may be fine, but you will find there are very few places like that in the United States or Canada. If more insulation isn't needed for warmth, it is needed for cooling.

The problem is the joists, their lower insulation value and their width. In a house where factory-built trusses were used for the roof, the lower chords, which are also the ceiling joists, are only 3½ inches high. Their average R-value is 3.5. If thicker insulation is used, such as 8-inch fiberglass batts (R-29), then there will be a strip between each batt, the joist, that is only R-3.5, or considerably underinsulated. There is also a greater chance of voids along the side of each joist.

To eliminate both problems as much as possible, it is better to put two layers of insulation in the attic. If placed from below, the first and thicker layer will be put on top of the ceiling joists and laid at right angles to them. In the shallow area over the eaves, the insulation will have to be cut to fit between the rafters to get the desired thickness. Consider using a rigid insulation with a higher insulation value per inch, or a special construction technique to handle that. Also be sure you don't block ventilation. (Vent/baffles should also be used with batt insulation.)

The top layer is installed across the ceiling joists, then another layer, the thickness of the joists, is put in from below. This method eliminates any underinsulated strips at the joists, gets rid of as many of the voids along the joists as anything except blown insulation, and prevents those discolored stripes found on the ceilings of inadequately insulated houses.

Underinsulated Strip

Another major reason for planning ahead is to avoid the built-in problem found in most standard construction. In most construction techniques used in this country there will be only 3 to 5 inches of space between the top of the wall and the bottom of the roof sheathing along the eaves. Without some forethought, this will leave a strip up to 3 feet wide along each eave that is underinsulated. Depending on the pitch of the roof, the square footage of this strip may total as much as that of a small room. While builders would not consider leaving the ceiling of one room underinsulated, they ignore the strip along the eaves, which can be just as large a space. This strip may have only one-third as much insulation as is required.

There are two ways this can be eliminated. One is to use one of the special construction techniques, shown in the diagrams, or to use an insulation with a much higher R-value than the normally used attic insulations. Either way, it should be planned for before construction begins.

If the builder opts for a higher value insulation, isocyanurate is the best. Only 5 inches is needed to attain an R-40 value whereas it would take over 18 inches of blown fiberglass or 11 inches of fiberglass batts.

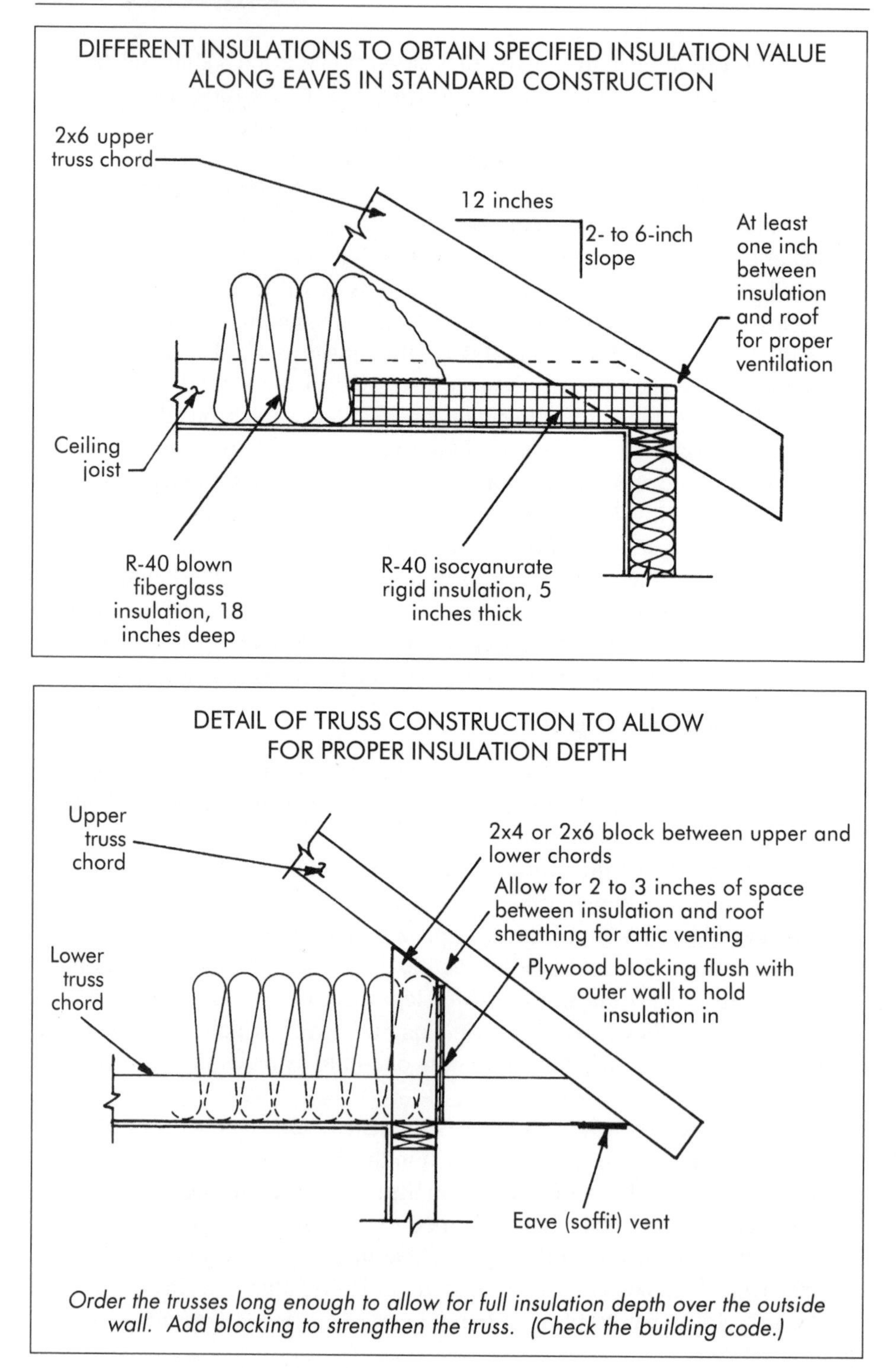

Order the trusses long enough to allow for full insulation depth over the outside wall. Add blocking to strengthen the truss. (Check the building code.)

Despite its higher cost, in that small area isocyanurate cannot be surpassed.

Longer Roof Trusses

One of the construction techniques used where a large space over the wall is needed is to have the roof trusses built longer than would normally be needed. This can give the needed height over the wall for blown insulation. Two things need to be done if this method is chosen. First, a support block must be added in the truss to strengthen it where it rests on the wall. (See diagram.) Second, and perhaps more important, check with your local building inspector to see if it is allowed. (Remember the comment about building codes working against good insulation practices?)

Another method used years ago was to build the ceiling as much as a foot below the top of the wall. The space between the ceiling and the top of the wall was filled with insulation. This method has the added advantage of placing a continuous layer of insulation from the inside of the wall completely across the ceiling. It has the disadvantage of costing more for construction, and in some instances, being against the local building codes. (See diagram for specifics.)

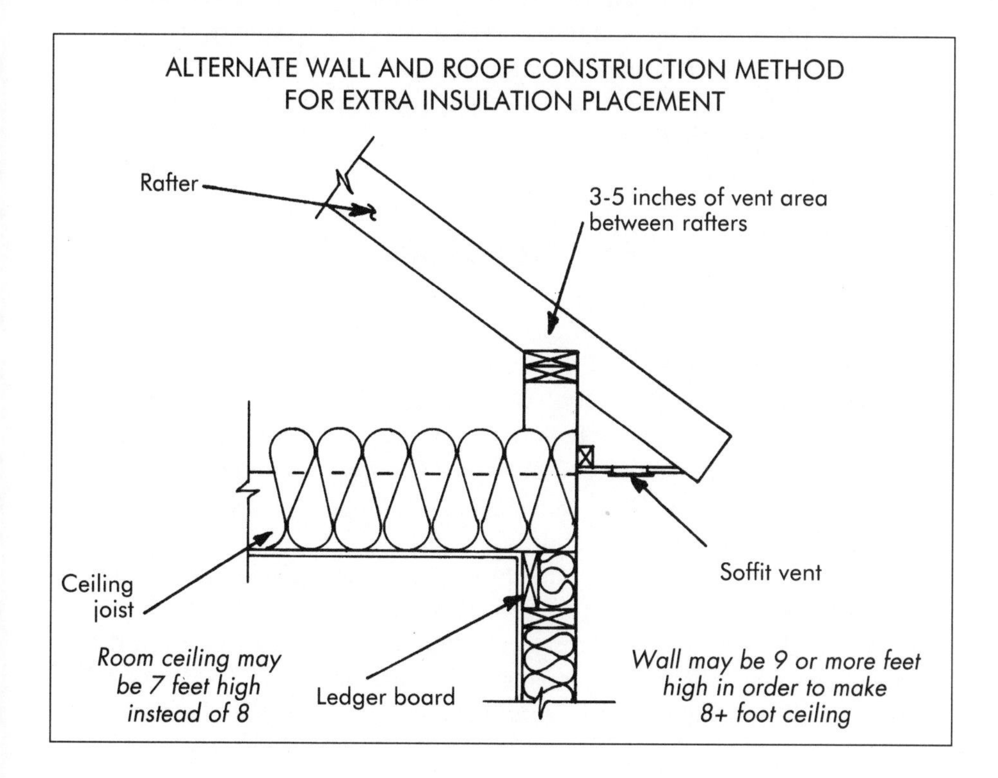

Fire Blocking

Some building codes work against good insulation practice. In some locales a solid layer of wood called fire-blocking is required at no less than 48-inch intervals in the outside walls of a structure. If the builder is trying to break the conduction of cold through the wall studs and attendant framing members, this lessens its effectiveness. This is why many people involved in insulation engineering are also pushing for major building code revisions in many parts of the country.

Avoid Pull-Down Stairway

Some homeowners try to utilize attic space, especially where the pitch of the roof provides headroom, for storage space. They may put a pull-down stairway kit in a hallway or one of the bedrooms. Insulating and installing a vapor barrier around the stairway is a nightmare, with even the best job leaving a lot to be desired. Any opening into an attic through the ceiling insulation and vapor barrier will cause problems with heat loss and moisture. If you plan to use the attic, insulate the roof instead of the ceiling.

Even if you live in a climate where a vapor barrier isn't required, you will have problems with ceiling insulation. Insulation is also used to keep heat from the attic out of the house. What would you want to store in a place that can easily reach 130°F on a sunny day? And, do you want a hole through the attic insulation that will allow heat into the rest of the house and waste your cooling energy dollars?

11. The Roof

SEVERAL ROOF DESIGNS require that insulation be placed directly in the roof instead of in the ceiling. Unfortunately, many builders simply stuff insulation between the rafters and let it go at that. Wood doesn't have the same insulation value of fiberglass or any other insulation, so there is the immediate problem of the rafters, plus possibly not enough space for adequate insulation between them.

Another problem that is ignored is ventilation. Even with a vapor barrier in place, some moisture is going to get into the roof. If the rafter space is stuffed full of insulation, with no ventilation of any kind, the homeowner may soon notice moisture spots on the interior sheathing of the house. If the vapor barrier is not installed, or if it leaks, the interior sheathing can be damaged to the point of pulling off the rafters.

Using Urethane

Some builders kill two birds with one stone by insulating the roof with urethane foam insulation. Urethane does act as a vapor barrier in this case, *if it is thick enough.* It has been found in some parts of Alaska that 6 inches of foam are required to do the job. This gives an R-30 value in the roof, which is not quite enough in colder areas.

Too, foam can shrink, especially if it isn't applied heavily enough. Builders in Alaska have also found that foam must be applied in at least 2 pounds per square foot weight or it will shrink and pull away from the rafters, leaving wide voids in the roof. (A house with this problem can easily be spotted when there is frost on the roof. The voids will show up as unfrosted strips, while the insulated areas will have frost on them.)

Space Needed

In any roof design calling for insulation either between the rafters or elsewhere in the roof, at least an inch of space should be left between the roof sheathing and the top of the insulation, and both top and bottom strip ventilation should be installed. This will insure adequate air movement under the roof sheathing. (As mentioned above, this isn't required with foam insulation if it is thick enough.)

To get the space above the insulation, it may be necessary to use a thinner insulation than would normally be used for the width of the rafters. Then, to get the amount of insulation desired for the climate, you may have to use wider rafters, or better still, purlins placed either below or above the rafters, with those filled with insulation. If the house will have a steel roof, the purlins can be placed on top of the rafters. They will give better support for the steel roofing material than rafters. (Most builders will use 2x4 purlins, placed flat, on the rafters to support steel roofing. If this is done, *do not* put insulation between the purlins. Then there will be adequate ventilation under the roofing.)

Use Two Layers

As has been pointed out, using two layers of insulation placed at right angles to each other is better than a single layer with wood between each strip of insulation. For story and a half house design, where there is a living space (usually a bedroom) directly under a roof, two or even three layers of insulation are better than trying to put all the insulation between the rafters. In this design it is imperative that there is adequate insulation

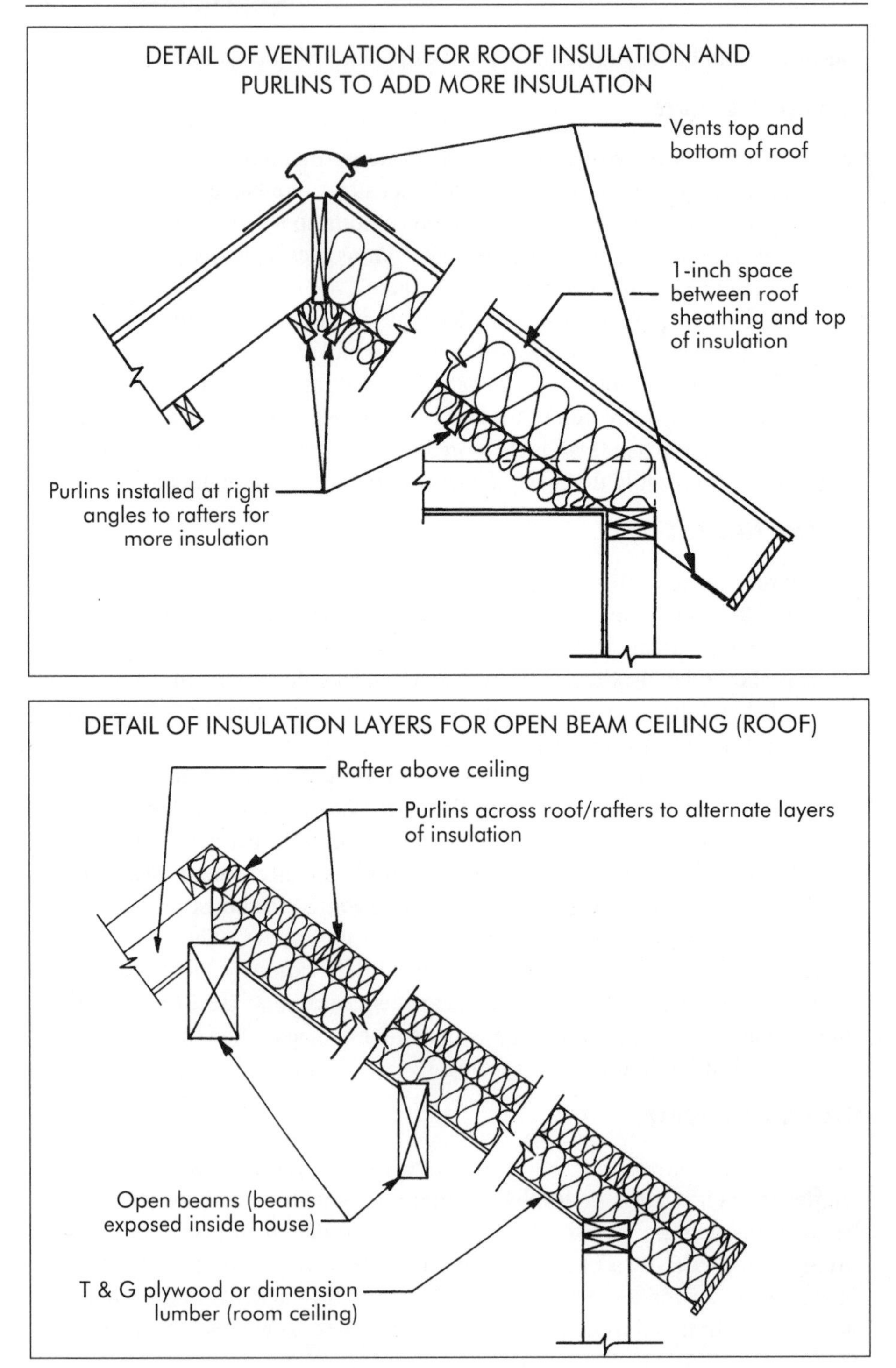
DETAIL OF VENTILATION FOR ROOF INSULATION AND PURLINS TO ADD MORE INSULATION
Vents top and bottom of roof
1-inch space between roof sheathing and top of insulation
Purlins installed at right angles to rafters for more insulation
DETAIL OF INSULATION LAYERS FOR OPEN BEAM CEILING (ROOF)
Rafter above ceiling
Purlins across roof/rafters to alternate layers of insulation
Open beams (beams exposed inside house)
T & G plywood or dimension lumber (room ceiling)

or the room will be cold in winter and hot in the summer and thus uncomfortable at least half of the year.

Another type of roof design requiring insulation directly in (or on) the roof is the open beam ceiling. This design uses heavy beams, which are decorative and hold the roof up. Depending on what the homeowner wants the ceiling to look like, the roof may be made with regular rafters, heavy (2 inches thick) planks, or tongue-and-grooved plywood placed directly on top of the beams. Whatever is used, the vapor barrier is installed on top of the first layer of material, under any rafters or any other type of roofing.

If rafters are used, it is advisable to place purlins on them to allow for two or more layers of insulation. Whenever this layered design is used, the rafters and purlins must be strongly fastened together to prevent any separation or movement from earth tremors or high winds. When a roof is built up using purlins and rafters above a beam ceiling, a high wind may lift part of the roof off if it isn't adequately fastened to the beams, ceiling, or rafters that are under it.

This problem is one inherent in some built-up roofs, where the builder simply glued one layer of rigid insulation to the roof, and subsequent layers to that, with the roofing material glued to the insulation. The roofing material most often used for this type of roof is hot mop, where a layer of hot tar is applied to the sheathing, then a layer of roofing paper, then a layer of hot tar, then paper, etc., to whatever thickness is desired. Two things are apparent. If installed over an open beam ceiling, there will be no vapor barrier on the warm side of the roof, and the final layers of paper and tar act as a vapor barrier, trapping moisture directly under them, which can bring about the moisture problems mentioned several times already.

Rigid Insulation

If the builder decides to use rigid insulation, especially the high R-value types, he still needs to use a rafter and purlin system to get the necessary thickness and cut down on voids at the edges of the insulation. Also, he needs to use a glue or caulking material on the edges of the pieces of insulation to further reduce any chance of voids and air leakage. Again, there should be an inch or so of space between the insulation and the roof sheathing to allow for air circulation and the removal of any moisture that may escape through the insulation.

Another Method

In some areas the problem of an adequately insulated roof has been approached from another angle. This is done by installing sandwich panels in place of a standard roof construction. These are factory-built preinsulated panels, usually 4 feet wide and 8 feet or more in length. They may be obtained in whatever length is required from some manufacturers.

Some of these are two pieces of plywood or particle board with a layer of rigid insulation between them and must be supported by rafters or beams. This also allows additional insulation to be placed, if needed. For these, the roof is framed in the standard way, and the panels then placed on top, with caulking put between them to prevent air leakage through the roof. Whatever roofing is desired is then put on top of the panels.

Foam Insulation Panels

Another panel system is also coming into use. The panels are framed with regular dimension lumber, then foam insulation is forced into them through holes in one end. They are made in whatever length the builder wants, then placed vertically on the walls and beams, with a caulking material between them to prevent air leaks. In climates where a vapor barrier is needed, both of these systems require it, even though some manufacturers claim the material used on the under side will act as a vapor barrier.

The cost of these panel roofing systems varies greatly, depending on the distance from the factory and the thickness of the insulation used. In some areas they may be worth the price because they reduce costs of framing material and labor. The homeowner will have to be very careful in figuring how much savings he may get from the use of panels as compared to a rafter and purlin roof.

Another method some builders have used to gain adequate insulation in a roof is to put a layer of urethane foam directly on top of a roof, then finishing it with either a spray-on or mop-on material. This takes almost yearly maintenance, as the urethane foam evaporates if left exposed directly to sunlight. The coating must be checked for breaks, such as where a broken limb may have hit it in a storm. These must be sealed immediately, or you may have a hole in your insulation. The coatings will also need to be renewed at much shorter intervals than most other roofs, and the coating material usually is quite expensive.

Again, planning on what is needed well before starting construction pays off with roof insulation. By deciding on the method of construction and the type of material that works best with that, homeowners can have a roof of the design they want, and with adequate insulation for their climate.

12. Soundproofing Walls & Floors

ALONG WITH KEEPING our houses warm or cool, insulation has another use. It restricts sound from certain areas. Bathrooms and bedrooms are two places we might want to restrict noise; a family room used by teenagers is another.

Too often no thought is given to this use of insulation until it is too late. Plan for it before construction begins. Many builders try to avoid it because it adds to their cost. While insulating for sound doesn't pay back in cash, it pays off in other ways just as tangible, such as peace and quiet.

Simply putting insulation into the interior walls as they are framed helps. But the materials used in constructing the wall, wood and gypsum board, are dense and readily transmit sound. In order to break this sound transmission through the solid part of the wall, it is necessary to build a double stud or offset stud wall. This wall will take more labor and material to construct, but for any room where you want to cut down on noise, it is worth it.

Plumbing Wall

This construction is often used in a plumbing wall — one containing pipes and therefore thicker than other walls. The simplest form is to use either 2x6 or 2x8 lumber for the top and bottom plates, and either 2x3 or 2x4

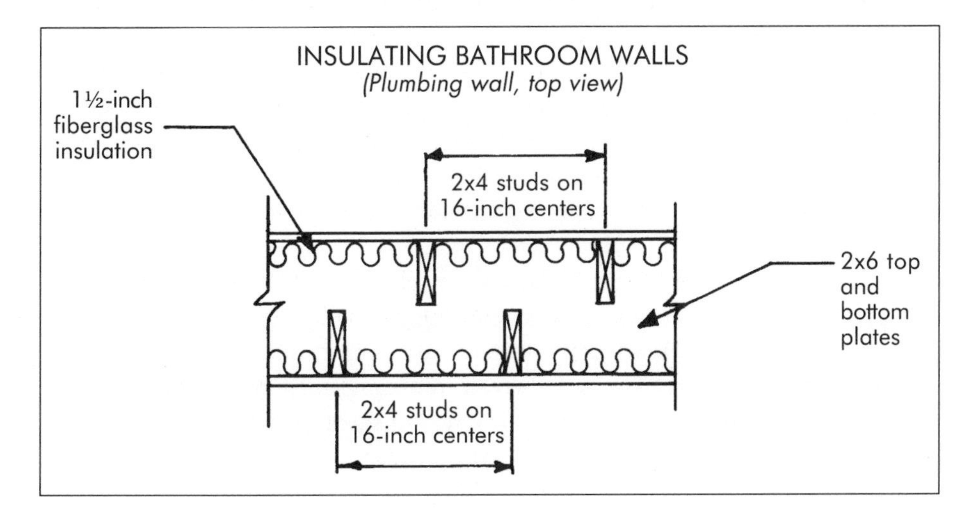

lumber for the studs.

The wall is framed the usual way, with one edge of the studs even with one edge of the plates. Then another set of studs is nailed into the wall, with an edge even with the other edge of the plates. This gives ample room for the plumber to work. It also allows room for insulation, on either one or both sides and often with an air space in the wall, which also helps deaden sound.

This same wall can be used for bedrooms or other rooms where sound needs to be restricted. The wall may also be constructed with 2x4 plates and 2x4 studs, but with the studs turned flat instead of on edge. This too breaks the transmission of sound through the dense studs. To save money, consider putting the studs on 24-inch centers, instead of the normal 16-inch, and using ⅝-inch gypsum board instead of the weaker ½-inch material.

Other Insulation Areas

Insulation benefits in other places in a bathroom. Put on the outside of the bottom and sides of a bathtub, it slows the cooling of the water as well as reducing the noise of water running into the tub. Some tubs come from the factory with polystyrene insulation glued to the bottoms, often just for support, but it also helps for heat and sound. Fiberglass may be stuffed behind the tub, or it may be foamed with polyurethane.

You can glue thin sheets of polystyrene insulation to the inside of a toilet tank to prevent condensation on the outside. Some toilets are sold with insulated tanks. Kits to do this are also available from many building and plumbing supply houses. Along with preventing condensation on the toilet tank, the insulation also cuts down on the noise of flowing water when it is filling.

Another possibility is to have your plumber install a mixing valve which mixes warm water with the tank water so it is at room temperature, to eliminate condensation. This adds an expense, since the water must be heated.

Floor Insulation

Insulation is used to block sound transmission in floors. Carpeting has long been the only answer to noise passing through a floor in an upstairs room, but now a Japanese firm is making a sound-insulated multi-layered flooring that it claims is as effective in reducing noise as carpeting, but looks like a regular wood floor. It is made in sheets 1 inch thick and is a combination of sound-absorbing material, foam rubber, plywood, and

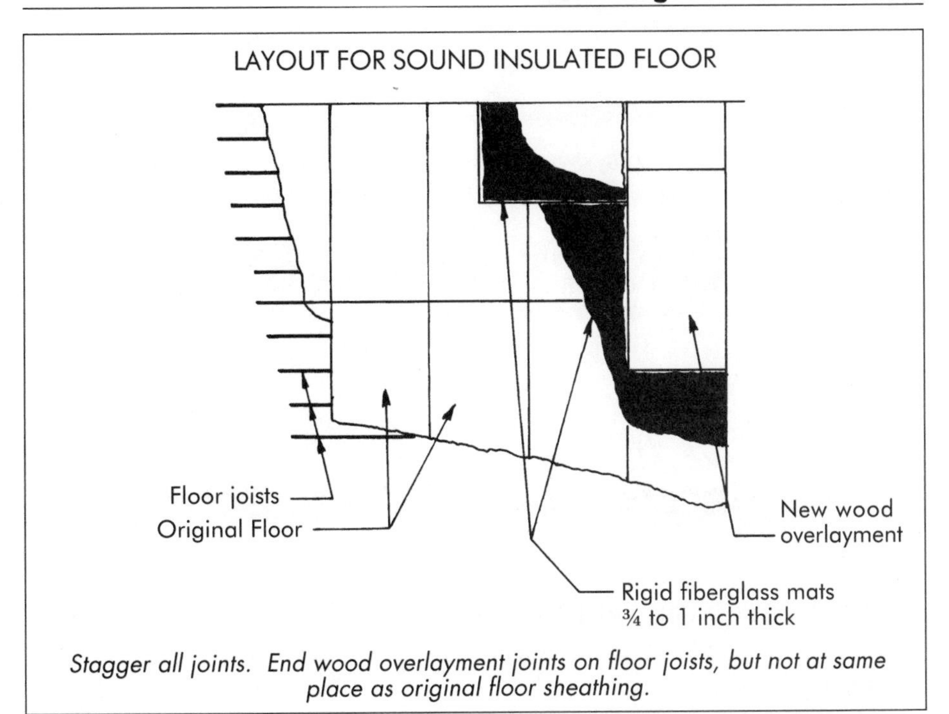

Stagger all joints. End wood overlayment joints on floor joists, but not at same place as original floor sheathing.

wood flooring. It is designed to be placed over another floor, including a concrete floor.

In Europe, builders glue rigid fiberglass insulation sheets to the existing wood floors, with sheets of plywood or particle board flooring material glued to that. Only the original floor is nailed to the joists. Glue is used to hold the upper two layers. This breaks any transmission of sound through the flooring. The top layer of flooring is not allowed to touch any of the surrounding walls, with a ⅛- to ¼-inch gap on all sides to prevent sound from traveling through the walls.

The rigid fiberglass sheets (it may be possible to use polystyrene, too) should be ¾ to 1 inch thick. They are laid with the long axis at right angles to the sheeting on the joists. The top layer of flooring is then laid with the long axis at right angles to the fiberglass, and parallel to the bottom layer of wood, but with the joints offset. The top layer of flooring should be tongue-and-grooved particle board so the edges won't sink into the insulation when walked on.

The glue may be either one of the panel adhesives put in with a caulking gun or regular carpenter's wood glue, which can be dribbled directly from the container. The edges of the fiberglass and of the top layer of wood should also have a bead of glue on them. Don't skimp on glue.

13. Basements and Crawl Spaces

UNLESS YOU ARE PREPARED to do some extensive excavation, insulating the foundation or basement walls of an older house will have to be done from the inside.

Before attempting this, check the walls for leaks. Leaky spots can be cracks in the concrete or blocks in a foundation wall, under the footing, or between the footing and the bottom of the wall. Any leaks must be sealed before any insulating is done.

Seal Cracks

Cracks in the walls must be sealed, preferably from the outside. This means digging down to the footing at the area of the crack and then working tar or plastic roofing compound into the crack and covering it and the adjacent area with black polyethylene.

Some cracks can be sealed from the inside, but the sealing material may work loose over time and there will be another leak.

Material used to seal cracks in a foundation wall should remain pliable even in cold weather. If a material hardens, it will either crack or pull away from the wall when cold and will not seal the crack.

Stopping Leaks

If a basement leaks only during a rain or perhaps spring runoff, you may be able to stop it without doing extensive excavation. Try removing about a foot of topsoil from an area 5 or 6 feet wide next to the problem wall. Slope the bottom of this area away from the wall, with a slope of 3 to 6 inches. Lay a layer of heavy polyethylene in this excavated area, bringing one side up the wall to a point above the finished grade, the level of the soil after the job is done. The plastic sheeting should be sealed to the wall with black plastic roofing compound so no water will run down the wall behind the plastic. Replace the soil in the area you have excavated, and slope it away from the wall with a drop of at least 3 inches in 3 feet. All the areas around a house should slope away from the foundation, whether or not you use this plastic sheeting system.

Choosing the Insulation

Once you know the basement or crawl space is dry, you can decide what insulation to use. In most cases, cost will be a deciding factor.

For a foundation wall, if you decide to use fiberglass, you need to put

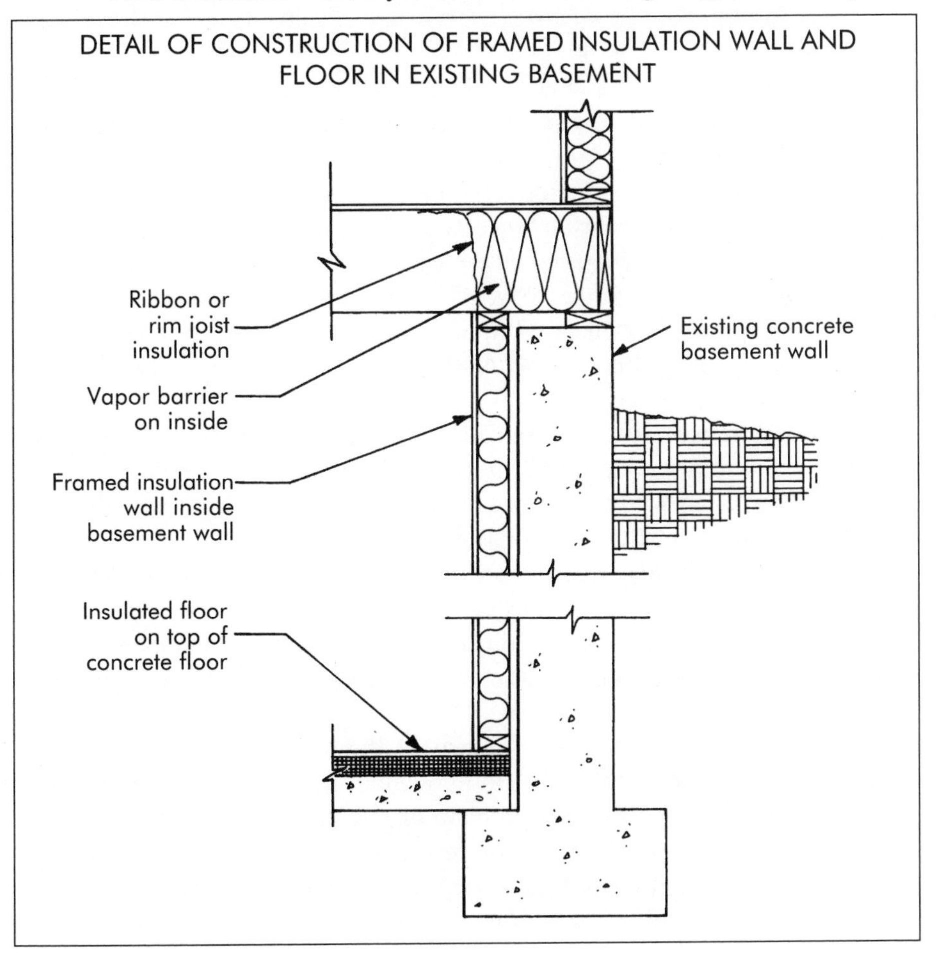

a vapor barrier against the concrete. This will prevent any moisture from coming through the wall from soaking the insulation. When you place the vapor barrier on the bottom of the crawl space, extend it enough above the bottom of the material on the wall so that any moisture that accumulates on it from the foundation wall will go into the soil instead of the atmosphere of the crawl space. No other vapor barrier should be needed in that space.

Insulate Rim Joist

The rim joist space, that space between the floor joists along the outside wall above the foundation, should be insulated first, then the vapor barrier brought up to the floor, so it will prevent moisture from accumulating in the rim joist insulation. The vapor barrier for the rim joist may be stapled or glued to the floor and the floor joists, or held in place with thin strips of wood. As with the vapor barrier everywhere else in a house, it should be as moistureproof as possible.

Foundation and basement walls as well as the rim joist space may also be insulated with any of the plastic products such as spray-on urethane, polyurethane, polystyrene, or isocyanurate glued directly to dry concrete. Do not put such products on a damp wall, or one that might get damp. Even though the insulation is waterproof, it may mildew and smell. A vapor barrier may be placed on the warm side of these rigid insulations as it is anywhere else in the house. The vapor barrier on the floor of the crawl space should extend well up over that on the wall, or it should be continuous from the floor above, all the way across the crawl space. The latter is not especially easy to do.

Some houses with dry foundation and basement walls have been insulated with spray-on cellulose fiber. This is probably the least expensive material to use, but it must be used in a dry area otherwise it will absorb moisture and eventually fall off the walls.

One of the primary reasons for placing a vapor barrier on the floor of a crawl space is to prevent moisture from the soil under the house from rising and entering the house, and particularly the framing members under the house. This is also the reason for providing vents in the walls of the crawl space.

It may seem to be a waste to insulate the foundation and then leave the vents open. Remember that in the winter, humidity goes down with the temperature, therefore the problem of moisture in the crawl space is reduced. Thus in most areas where it can get cold enough to freeze plumbing in the crawl space, the vents may be closed. (Be sure not to close any appliance's combustion air vent, though.) Some heating sys-

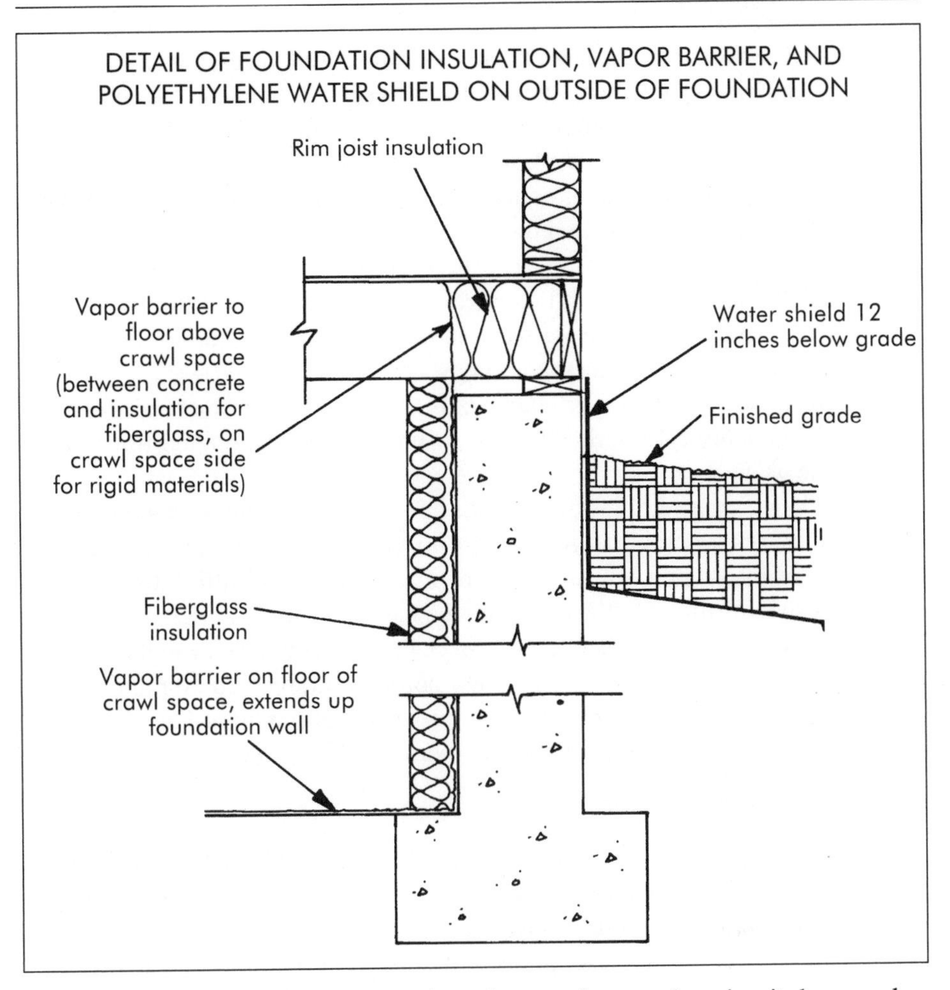

tems put a small amount of heat into the crawl space, keeping it dryer and incidentially warming the floor of the house.

There is some argument about closing vents in warmer areas, where temperatures get below freezing only once in ten years. Since the humidity is higher, more ventilation is needed. In that case, the plumbing should be insulated against that once-in-ten-years freeze.

Insulate Heat Ducts

In such an area, heat ducts in the crawl space should also be insulated. No one wants a blast of cold air from the heating system, even if it isn't freezing outside.

Of course, one way to eliminate these ventilation problems is to insulate the floor, and let the crawl space be vented outside. Insulating floors directly is covered in another chapter, as is insulating heat ducts.

Insulating a basement, especially for use as living space, can entail building a house within a house, by building a frame wall inside the basement concrete wall. The simplest way to do this is to frame a wall several inches inside the concrete, using 2x4 lumber. This wall should be far enough away from the concrete so the insulation will not touch the concrete and draw moisture from it. The base plate for this wall should be made from all-weather-wood (treated wood) to prevent rotting. A vapor barrier is installed in the usual way, on the warm side (inside) of the wall. It is finished whatever way the homeowner wants.

Contractors sometimes attach 2x4s to the walls by "shooting" boards onto the concrete. Then they fill the spaces with insulation and finish the wall the usual way.

There are several problems with this. First, shooting is done with a tool that uses a blank pistol cartridge to propel a special nail through a board and into concrete. As concrete gets older it gets brittle. Instead of being driven firmly into the concrete, the nail will simply knock out a chip and not hold anything. These chips can be dangerous to anyone in the area.

Second, concrete walls are often not very straight, so attempting to make framing lumber conform to them is nearly impossible. This is also a reason why gluing rigid insulation to them often isn't successful.

Concrete Block Walls

Concrete block walls have the same problems as solid concrete, with the exception that they may be straighter and thus may be more readily insulated with glued-on rigid insulations.

In some areas builders glue rigid insulation onto block walls, then glue gypsum board or paneling to the insulation, eliminating any framing. It is essential that the block wall be completely dry for this type of job.

In an older house, it may be necessary to dig a trench around the outside of the entire basement wall, then cover the outside of the wall with waterproof material to keep the wall dry enough to glue on insulation.

Insulate and vapor barrier the rim joist above a basement wall before doing any ceiling finish work. This is a potential trouble spot in every house.

When insulating the basement in an older house, most homeowners will ignore the floor. Unless it is so bad it must have a new layer of concrete poured on it, about all that will be done is to carpet it and hope that is enough to keep the user's feet warm.

No matter what the condition of the floor is, it can be insulated without pouring another floor. (If another floor must be poured, use a vapor barrier and insulate it as mentioned in the chapter on basement floors.)

To insulate an existing concrete floor, seal all cracks with a non-hardening sealer as for radon then seal it with concrete sealer to prevent the intrusion of moisture. If there is any question about moisture, a vapor barrier of polyethylene film or similar should be put down directly on the floor.

Insulate Under Floor

A wooden floor is then framed on top of the concrete, with insulation between the floor joists. The joists should be 12 to 16 inches apart and deep enough to allow needed insulation as determined by the formula in chapter 4.

If the wood is put directly onto the concrete, all-weather wood should be used for the framing members. If the floor is supported, the supporting blocks should be of all-weather wood. Putting carpet directly on a concrete floor as a way to insulate it is considered an inexpensive solution. If the floor will remain totally dry, it will provide some insulation, but the homeowner should consider many variables. First, the concrete must be sealed with a special sealer. The glue used for this kind of application is more expensive. The carpet itself must be designed for that use. Not every glue-down carpet will work on concrete, especially if the floor is below ground level. If extreme care isn't taken to prevent moisture from getting to the carpet, even the best material may soon be a smelly, soggy mess.

While insulating a basement isn't the easiest or least inexpensive job a homeowner can do, insulation will pay for itself here. Extra living space, warmer upstairs floors, and reduced energy costs are all benefits from doing an insulation job in an existing basement.

14. Exterior Walls

INSULATING, REINSULATING, or adding insulation to the outside walls of an older house can be a challenge.

Houses constructed before 1940 often have an extra layer of lumber in the walls, on both the inside and the outside. This makes for a very solid wall. But unless it was insulated when built, it can be a trial for someone trying to add insulation to the house.

A quick fix which has been used all over the country is to pump the walls full of urea-tripolymer foam.

This is done by boring holes 2½ to 3 inches in diameter between each pair of studs. One hole is bored at the bottom, just above the bottom wall plate, and the other at the top, just below the top plate.

More work is required if the wall has fire blocking, which is solid blocks placed between the studs and about every 48 inches up the wall, to prevent the spread of fire through the walls. In this case additional holes are required above and below the fire block.

The foam nozzle is inserted in the bottom hole and foam is forced in until it comes out the top hole.

Use of Urea-formaldehyde

At first, urea-formaldehyde was used extensively for insulating this way. It was dropped when fear spread that the health of those in the home was adversely affected by fumes from the formaldehyde.

Many persons who had used the material removed it because of this possible threat to their health. They did this by removing either the inside or, less commonly, the outside wall sheathing, then carving out the dried foam. At this time, many found uninsulated gaps in the walls, usually in the top corners.

Some recently published articles indicate that the Environmental Protection Agency has discovered the hazard with the use of urea-formaldehyde foam is less than first believed. The formaldehyde dissipates quickly after the foam dries.

If your house has been insulated with this material, have tests done to determine the level of formaldehyde in the atmosphere of the house. Your local EPA or Home Extension Office should be able to do these tests or tell you who can. If no danger is found, this can save you the expense of tearing out the inside wall sheathing, carving out the insulation, and reinsulating.

Another Method

Another way such a well-constructed wall may be insulated is to remove the top two siding boards (assuming the house has lap siding) and cut a strip out of the sub-siding to expose the openings between the studs just below the top plate. Then you can pour or blow (with low pressure and very carefully) expanded mica insulation into the spaces in the wall. Once the space is filled as full as possible, compress a piece of unbacked fiberglass insulation into the remaining space so that as it expands, it will compensate for any settling of the mica. The boards that were cut out can be put back, or replaced with a board of the same thickness, and the siding put back on to cover the opening.

As with foaming in insulation, fire blocking and areas under windows cause problems. To insulate under windows and below fire blocking entails removing a strip of siding under each one, and cutting into the wall as is done at the top. The insulation is then poured in, and a piece of fiberglass inserted to fill the void left and take up any settling of the mica.

Some old houses have an opening from the basement to the attic through the walls, between the studs. You can drop a marble in the attic of one of these houses and it will roll out in the basement.

That was the problem with them. A small fire in the basement became a roaring inferno going up the walls to the attic. This is one reason for the use of fire blocking in the walls of later construction. There are still a lot of old houses in the Midwest and the East that were built this way and aren't insulated, or only have insulation in the attics. A house built this way may also be foamed with ureatyrpolymer, either from the basement or the attic. This may present problems because most of these houses are at least two stories high with an attic, and the foam may not completely fill the wall space.

To insulate such a house with poured insulation (mica), you must block the openings into the basement. Use a solid material such as dimension lumber or plywood, and fasten it securely into the wall so the weight of the insulation won't push it off and allow the insulation to leak out of the wall.

Because of the way most of these houses are built, with the walls being a foot or more above the ceiling, the wall insulation can easily merge with the attic insulation. This is a benefit because there will not be a gap in the insulation where the top double plate of a wall normally is.

Vapor Barrier Needed

A major problem in insulating an existing house is how to put a vapor barrier in place. In the cases just discussed, you don't, without some more work and expense. Some studies have concluded that urea-based foams will themselves act as a vapor barrier. This is true to a certain extent. If the foam is thick enough, it does stop problems with condensation in roofs. From local observations, however, it seems this is in the 5-6-inch range, and not many house walls are that thick. In a very cold climate, where moisture may be a problem, the installation of a vapor barrier should be considered. Perhaps the homeowner may wait a year after insulating the walls, making careful observations of the house walls to detect any moisture problems before making a decision one way or the other.

The interior sheathing of many homes may be so damaged that it might be advisable to put new material on the walls. Gypsum board or paneling may be used. If this is done, a vapor barrier can be installed under it, between the old and the new material. Outlet boxes and window and door frames should be treated the same way they are in new construction, with every avenue of air leakage sealed with caulking.

If the house is constructed the way most have been since World War II, the owner may want to take care of two problems at the same time. The interior wall sheathing in these houses is often only ½-inch gypsum board, and over the years has become misshapen, damaged, and discolored to the point where paint won't cover the spots anymore. In that type of house, it may be less costly to remove the interior wall sheathing, insulate the wall as for new construction, put in a vapor barrier, and refinish the wall with new gypsum board or paneling to suit the owner's tastes. If this is done, the electrical wiring should be checked and replaced if necessary at the same time.

Rigid Insulation

If none of the methods described can be used, or if they are too costly (remember, insulation pays for itself), isocyanurate, polyurethane or polystyrene rigid insulation may be put on the outside, inside, or both of the walls. This may seem to be an easy way to go, but it also entails adding to the window frames and extending outlet and switch boxes, plus putting on either new siding or new interior sheathing.

There is "insulated siding" on the market, but a close look at the R-value chart will show these may not give much insulation to your house. One form is made of the same material used for roofing, like asphalt shingles only thicker. Even if 1 inch thick, this doesn't add much insulation value to a wall. It also looks like roofing.

Some of the metal and plastic siding materials have a thin layer of either polystyrene or cellulose fiberboard (¼ to ⅜-inch) that fits in the space under each piece as it is put on the house. This has been called insulated siding, but the insulation value is very low. Because of the lap design of the siding, the "insulation" does not cover the entire wall, so every 8 to 12 inches there is an uninsulated strip of wall. Again, because of the propensity of some persons to cover moisture problems with this type of siding, anyone contemplating buying an older house which has been renovated using that material should be very careful.

15. Ceilings and Attics

THE ATTIC OF AN OLDER HOUSE is often all that gets insulated. It is the easiest to get at, and is where most of the heat is lost. Some engineers estimate as much as 85 percent of the heat loss in a house is through the attic. Since it is so easy to do, too many homeowners simply have a lot of insulation blown into the attic and let it go at that.

In such hit and run insulating, two things are generally overlooked: ventilation and a vapor barrier. The vapor barrier can be eliminated if there is adequate ventilation. Again, as in new construction, piling in insulation sometimes blocks the ventilation and the problems are compounded by the lack of a vapor barrier.

Ventilation

Most government reinsulation programs stress having adequate ventilation before putting more insulation into the attic of an older house. This can work into a job. Most houses built since the 1950s have some insulation in the attic, but many rely on the many small holes in construction for ventilation. Shingle roofs, for example, have a lot of little holes, but not enough for adequate ventilation of an attic with no vapor barrier.

Before insulating, check for vents in the soffits, gables, ridge, and other places in the roof. Remember the formula for venting: If the ridge or roof vents are 3 feet or higher above the soffit (or lower roof) vents, the total net free area of all vents should be at least 1 square foot for every 300 square feet of ceiling. If the roof vents are less than 3 feet higher than the soffit or lower roof vents, the minimum amount of net free area of all vents is 1 square foot for every 150 square feet of ceiling. When in doubt, put in more!

Gable Vents

The simplest vents to put in are gable vents. These should be put into every gable on the house. Many styles of ready-made gable vents are available from building supply stores, or you may decide to build your own. Whatever you use, the vent should have baffles to prevent the wind from blowing the insulation around and keep wind-driven rain out, as well as screening to prevent the intrusion of insects, birds, and bats. In some houses, a large vent in each gable will be enough ventilation for the attic.

Remember that "net free area" isn't the same as the area of the vent. Subtract the widths of the material used for baffles, and reduce your answer by a little more to compensate for the screen.

If the house has a truss-supported roof, a continuous ridge vent can be installed without a lot of work. This vent is made of aluminum, and is designed to replace the ridge shingles on a house roofed with asphalt shingles, or the ridge cap on some metal roofs. To install it, remove the ridge shingles or ridge cap, and cut a gap in the plywood roof sheathing the length of the ridge. Screw or nail the ridge vent over this gap, using a heavy bead of plastic roofing compound to seal the edges to the shingles. Two methods are possible with a metal roof. The edges can be sealed with the rubber-like strips designed for sealing that roofing. Or since the regular ridge was removed, that ridge can be split in two and each edge used to make a seal along the edge of the ridge vent. There should be an equal amount of soffit vent area for proper air flow.

Roof Vent

To install a roof vent, a hole must be made in the roof. As these vents were designed to be put in as the roof was being shingled, it takes a little effort to put them in later. Many asphalt shingle roofs have some type of tab sealer on each shingle. If the weather is too warm, this sealer will cause the shingle to tear and possibly make a hole where one isn't wanted. A cool morning, or even a cold day is better for installing such a vent. (I have put these vents in during below-zero weather, and definitely do not recommend that, either.)

Remove just enough of the shingles to slip the flange of the vent under them to make a good seal. Once an opening has been made in the shingles, cut out a piece of the roof sheathing at least as large as the opening of the vent. Put a bead of non-hardening caulking (plastic roofing compound or silicone-based material) across the top of the vent flange and down the sides as far as the shingles cover. Put another bead on the roofing so that the underside of the vent flange will be sealed against the roofing, down both sides and across the bottom. Slide the vent into position and nail or screw the flange to the roof in at least four places, preferably under the shingle tabs. The same method may be used for a wood shingle or shake roof.

To put a vent on a roof that is made of layers of roofing paper and tar, cut a hole in the roof the size of the vent opening, and glue the vent to the roof with copious amounts of plastic roofing compound or hot tar, then nail or screw the vent solidly onto the material. Take care to assure there are no leaks.

Normally roof vents will be installed near the top of the roof. This helps keep them from leaking, as there is less water at the top than farther down the roof. However, it may be necessary to install a vent lower down on a roof where it is impossible to gain venting from the eaves or soffits. Sealing against leaks is most important on any vent, especially a lower one. The idea is to vent moisture out of the attic, not let it in.

Soffit Vents

Putting in soffit vents may seem like a simple job, but making them work is another problem. The problem is caused by insulators piling insulation against the bottom of the roof sheathing and blocking the air flow between the soffit and roof vents.

There are several ways to eliminate this problem. One is to crawl into the attic and put in the insulation baffle/vents described in the chapter on new construction. If there is no insulation in the attic, this isn't too hard to do. If there is some insulation, then it is more work. First, the insulation must be raked away from the top of the wall, so it won't interfere with installing the baffle/vent.

Lay boards across the ceiling joists to support yourself as you put in the vents, and prevent you from falling through the ceiling. Then staple the vents in place. (Do this job on a cool to cold day, as it can get warm in an attic.)

Some builders make a tube 3 or 4 inches in diameter and insert it into the opening between the top of the wall and the roof sheathing. This allows air to enter the attic from the soffit. It isn't as good as the baffle/vents, but is easier. Use felt paper rolled into a tube, or something like dryer vent tubing, either aluminum or plastic. Make sure both ends are clear of insulation so the air may flow freely through it. Some building codes and government-financed re-insulation programs prohibit this method of venting an attic, so check that before going ahead with it.

Best Method

Probably the best method of getting proper attic venting from the soffit and getting an adequate amount of insulation under the eaves is explained in the chapter on new construction. That is to use a high R-value insulation in the first 2 or 3 feet in from the eaves. This will keep blown insulation from piling against the bottom of the roof and blocking air flow, and give the proper amount of insulation along the eaves of a house at the same

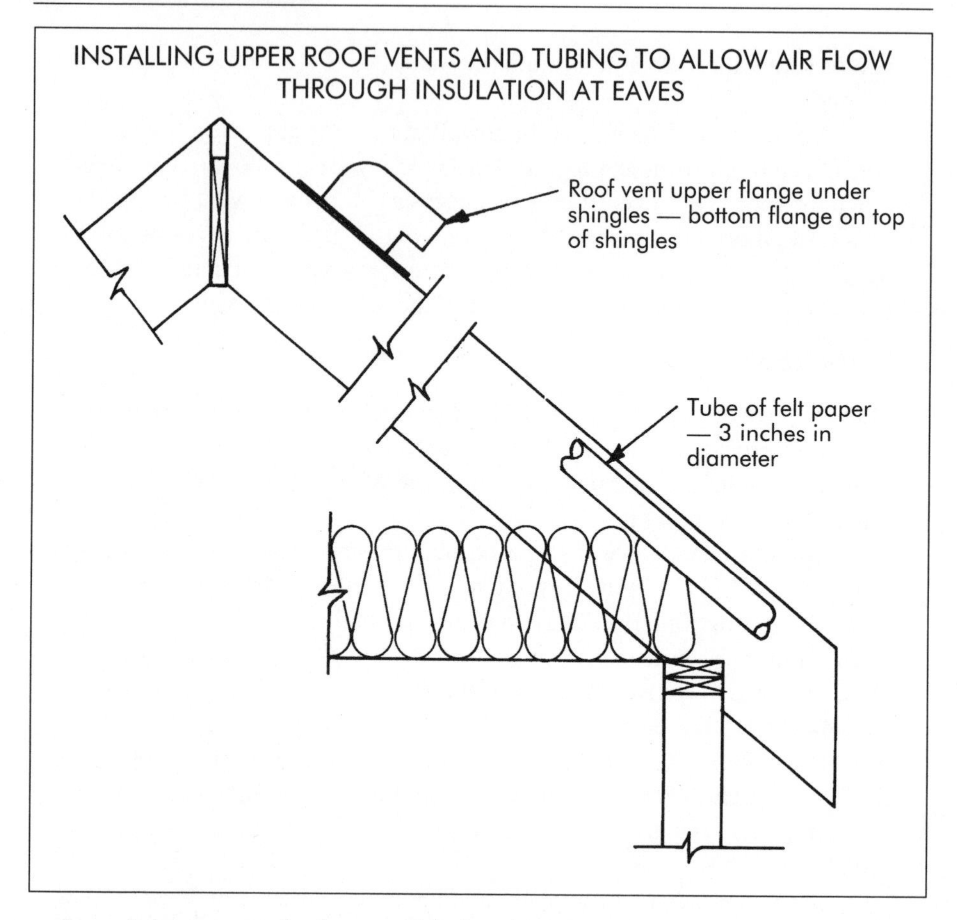

time. It is more costly, but considering the labor involved and the fact that insulation pays for itself, it is probably the best way to go.

Dams Needed

Before adding insulation in the attic of an older house, install dams around recessed light fixtures, chimneys, and the range vent pipe. This may entail digging out some old insulation, but it is necessary. The local building code or regulations with a government loan for insulation work may require this.

All bathroom fans vented into the attic must be extended to the outside. If taken through a roof, they should have their own roof vents. Some contractors in the past have simply brought the pipes into a regular roof vent and left it that way. This is a violation of most building codes. The range hood especially should be vented through its own vent outside. One vented into the attic is a definite fire hazard.

Electrical Problems

You may find problems in the electrical wiring in many older houses. If the house has ceiling-mounted light fixtures, the wiring in the mounting box may be cooked by heat from the fixture, and need to be replaced. If the fixture has a large glass globe enclosing the lights, or if it has three or more light bulbs, it should be removed and the wiring checked for broken or brittle insulation.

In any house that lacks insulation in the attic, all the wiring should be checked for damage, and replaced if necessary before insulation is put in the attic.

Vapor Barrier

Putting a vapor barrier in the attic of an older house can range from laying it across the joists and the existing ceiling to putting it on the underside of the ceiling to putting in a false ceiling. In short, from a little effort to a lot of effort.

If the house has never had any insulation and the ceiling joists and adjoining framing members are reasonably flat without a lot of sharp points or edges, lay polyethylene material directly on top of them, with enough slack to allow it to sag against the ceiling when the insulation is put on top of it. Allow ample material to overlap all joints so they won't leak moisture. Use silicon sealer so the joints won't leak.

If a vapor barrier is necessary in your climate, and your ceiling already has some insulation or the vapor barrier can't be put in on top, it may be necessary to put polyethylene on the underside of the ceiling, then refinish the ceiling with a new covering. If the original ceiling material is water-spotted or damaged, as is often found in older houses, this may be the best method.

Many houses constructed before 1940 have ceilings up to 12 feet high. The primary reason for this was cooling in summer, as the high ceilings kept the heat above the people. This wasn't such a bad idea before air conditioning and houses that are heated evenly throughout. Now, a 12-foot ceiling requires you to heat or cool an extra 4000 cubic feet of space for every 1000 square feet of living area.

Installing a false ceiling below the original is another way to put both a vapor barrier and adequate insulation into an attic. It is expensive initially, but again, insulation pays for itself. (I grew up in a house with 12-foot ceilings, with absolutely no insulation in the attic. It was uncomfortable most of the year.)

16. The Roof

INSULATING THE ROOF of an older house can tax the ingenuity of the most dedicated do-it-yourselfer. Many old houses have steep roofs. As families grew, owners through the years often built rooms in the attic. Insulation wasn't considered essential, as most homeowners just heated the living space and piled on more blankets in the bedrooms.

You may find a layer of boards nailed diagonally on the bottom of the rafters. This provides excellent bracing for the roof, but raises havoc with any plans to reinsulate.

Even if you put in a vapor barrier, you must provide ventilation, as in new construction of this type. The easiest way would be to foam the rafter spaces with urea-tripolymer and let it go at that. If you do, use at least 2 pounds per square foot, or you will have shrinkage problems. This material can take care of any moisture problems if a good vapor barrier is installed on the warm side of it.

However, in most construction the rafters aren't wide enough to put the required R-value of insulation in the spaces between them. There is also the problem of the rafters themselves. Since they have a much lower R-value than the insulation, they can conduct cold into the room under the roof. It may be necessary to foam the rafter spaces, then add a layer of rigid insulation on the bottom of the roof and refinish it with a vapor barrier and gypsum board or other material.

Use Purlins

Another way to add insulation is to place purlins on the bottom of the roof, then insulate between them with fiberglass, and finish the ceiling with a vapor barrier and gypsum board.

If the underside of the roof is accessible, it may be insulated with fiberglass, leaving at least 1 inch of space between the insulation and the bottom of the roof sheathing for ventilation. (See illustration in chapter 11.) Be sure to provide openings at the top and bottom for free air flow.

There probably won't be enough space to put in the desired thickness of insulation so purlins will have to be added at right angles to the rafters to get the proper depth. Be sure to fasten the purlins to the rafters. Try the steel "hurricane ties" used to fasten rafters or trusses to a wall.

A fix-all solution has been used in some places by foaming a layer of urethane insulation to the outside of an old roof. This is claimed to give insulation and seal the roof against leaks. It requires maintenance, with a fresh coat of sealer every year or so, depending on the climate, or you will see your expensive insulation evaporate when the sun shines. Any damage to it, such as from falling limbs, must be patched as soon as possible or a hole may "evaporate" into it.

Buildup of Roof

Another way to add insulation to an older roof is to build it up with polyurethane, polystyrene, or isocyanurate rigid insulation. If the roof sheathing is solid and will hold screws, this may be the way to go. By putting a couple of layers of rigid insulation directly on the roof, then covering that with steel roofing or another layer of plywood, and finishing with whatever roofing material is desired, you can add R-value to your roof and get a new roof in the bargain.

It may be necessary to remove all the old roofing material or to repair the old roof sheathing in spots to ensure that the fasteners will hold the new roof. (The insulation material should be glued to the roof sheathing, too.) Use screws long enough to reach through the insulation into the original rafters to fasten new sheathing or roofing. This puts a limit of about 3 inches on the thickness of the insulation unless purlins are placed on the old roof and the insulation put between them, as described in the chapter on new roof construction.

Provide Ventilation

You should provide some ventilation directly under the roofing to prevent problems from moisture buildup. Steel roofing material has crimps (grooves) in it which stiffen it and also act as channels for air to travel through. Other roofing on plywood sheathing will require air space under the plywood. Be sure there are adequate openings at the top and bottom of the roof to allow a free flow of air to remove any accumulation of moisture.

Insulating the roof of an older house can be a challenge, but as with all insulation, it will pay for itself over time. Considerable ingenuity may be required to put the desired amount of insulation in a roof.

Insulation between the rafters usually won't be adequate, so you will need to put more insulation on either the outside or the inside of the roof. If a new roof is needed, the insulation can be put on the outside, under the new roof, and two jobs done at the same time. If the extra insulation is put on the underside of the roof, there will be a slight loss in living space, but the trade-off in energy savings will be worth it.

17. Doors and Windows

WHAT DO DOORS AND WINDOWS have to do with insulation? They're in the outside walls of your house, and take space that could be filled with insulation. As such, they are one of the major factors in determining what the average R-value of your outside walls will be. For that reason, doors and windows should have as high an R-value as possible, and shouldn't take up more wall space than is absolutely necessary.

When designing a home in cold country, engineers use a rough rule of thumb: The total window area of a house shouldn't exceed 8-10 percent of the total floor area of the building. For instance, if a bedroom has 150 sq. ft. of area, the area of the window shouldn't be over 15 sq. ft. That can be a 3 x 5-foot window, which is a good size for a bedroom.

Move to the living room and this changes. For a living room with 600 sq. ft. of floor space, the rule says there shouldn't be over 60 sq. ft. of window area. That's not enough for most people, especially if there is a view.

Aim for High Value

Because they want to get the highest average R-value for the outside walls, most builders try to use windows and doors having the highest possible R-values. In most areas where cold weather is a consideration, the windows will be double glazed, which means they will have two layers of glass in each window. Some of these are made as a unit, with two panes of glass bonded to a metal strip, with from ¼ to 1 inch of air space between them. In extremely cold areas some builders put in two of these double-glazed units, with a space 2 to 4 inches wide between them. Alaska University engineers, who first tried this, believe that four panes of glass is the most that can be used in a window. The gain in R-value by using more is negligible and the cost is increased.

Double-glazed windows are made in three styles. The most efficient, and what most builders think of when they hear the term double glazed, is two panes bonded to a metal strip. Since they are sealed, no moisture or dirt can get into them. They will usually stay clear, no matter what the weather is. If the humidity in the house is too high, they will have some condensation along the bottom. Lower the humidity in the house and that will disappear.

To clean them, wash the outside of each sheet of glass. If moisture and dirt get on the inside, between the panes, they will have to be replaced. They may be ordered in virtually any shape or size, with or without any framing. For windows that can't be opened, many builders buy the glass, then build them into the wall. Otherwise, they can be ordered in a finished frame, either wood, metal or plastic, which can be inserted in a rough opening in the wall.

Removable Windows

Another type of readily available double-glazed window is the RDG (removable double glass) window. One pane of glass is permanently fixed in a frame, with a second pane mounted in a thin metal frame and held within an inch of the other by metal or plastic clips. The RDG system does not have quite as high an R-value as the sealed units, but the price isn't as high either. One drawback is that dirt and moisture can get between the panes. To clean, you must take the detachable pane off and wash both sides of both panes. This will have to be done at least twice a year in most places.

A double-glazed window that has been in use for many years is the storm window. A completely glazed window is mounted outside the main window, usually within 3 or 4 inches. In the summer it is removed and replaced with a framed screen. In the fall, the screens are removed and the storm window put back up, after all windows are cleaned. These involve a lot of work, without any extra benefits in insulation value, so are used less and less frequently.

A variation on the storm window is being used in Canada and Alaska. Removable panes are framed with thin metal, as the removable pane of an RDG window, then placed on either or both the inside and outside of a regular double glazed window during the cold months. They give a low-cost help during extremely cold times, when the temperature goes down to -40°F. or colder. They are a lot of extra work, but only a one-time expense.

Insulated Shutter

Another cold country solution to avoid heat loss through windows in the coldest part of the winter is used extensively in northern Canada and Alaska. This is a shutter made of plywood and polystryene rigid insulation 1 to 2 inches thick. The polystyrene is glued to the plywood, and the whole thing fitted over the outside of the regular window with the plywood out. This reduces heat loss through the window by as much as 50 percent. It also reduces the light entering the window by 100 percent. However, in a climate where there is little daylight during the winter, the loss of light is negligible while the heat loss is quite noticeable.

Some windows available (for a price) have an insulating medium that is removed or replaced automatically, as the temperature varies. I have never seen these units and have some doubts about their cost effectiveness.

There are several add-on items that can reduce heat loss through windows without the expense of extra glass or framing. One of these is a sheet of clear plastic which is taped to the inside frame of a window in early winter and removed in early spring. This material can be shrunk by using a hair dryer blowing warm air on any wrinkles after it is taped to the window frame. Variations of this can be made for pennies with polyethylene sheeting, but it isn't as clear and can't be fitted as well.

Drapes

Drapes, in all their variations, also help to reduce heat loss through windows. Even the simplest, hanging from a rod in front of the window, will have some effect. Several variations of drapes are made to add insulation to windows. One of these has tracks inside the window frame, so when the drape is closed, there is very little air movement near the glass. Another has magnets sewn into the material, and metal strips placed along the sides of the window, which also provides an effective seal. The term "window quilts" is used in some of the literature on these, which gives an idea of what they are designed to do.

Window frames are made of wood, plastic, or metal. For home use, the metal used most is aluminum, and steel is used in some industrial applications. Aluminum has no R-value and conducts cold readily. While a double-glazed window mounted in an aluminum frame will stay clear in freezing weather, the moisture in the air in the house will condense and freeze on the aluminum, causing problems with water stains and mildew around the window.

Wooden Frames

The most commonly used framing material is wood. It is attractive, may be stained or painted whatever color is wanted, is durable (within limits), and reasonably priced. In humid climates, though, unless it has been treated with a water seal and fungicide, it may rot out in as little as five years. When buying wood window or door frames, check the literature or ask the salesperson about this. Most major manufacturers of wood framed windows treat the framing material with some type of water seal. You can do this work yourself, but it is a messy and uncomfortable job, and the material used is somewhat toxic.

The "new kids on the block" are the vinyl framed windows. Manufacturers claim they have a higher R-value than wood and wear much better than either wood or metal. They also cost more. The vinyl is not supposed to become brittle in extreme cold, or shrink or expand with temperature variations. The windows are available in colors, but it may be difficult to change colors at some later date without using a special paint.

Patio door systems are available in the same frame materials as these windows.

Variety of Doors

A variety of exterior doors and their frames are made from wood and metal. There are wood doors in wood frames, metal doors in metal frames, and wood in metal and metal in wood. With the exception of panel doors, which have a thick wood frame around thin wood panels, exterior doors are uniformly 1¾ inches thick. A solid wood or a "solid core" wood door has an R-value of approximately 1.75, not much different than a double-glazed window.

Steel doors sometimes are hollow (definitely not something to use in cold country) but are usually filled with either polystyrene or polyurethane. They could have an R-value as high as 7, but with the steel covering this is reduced. The best insulated steel doors have a wooden edge between the steel sides to break any cold transmission from the outside to the inside. They also have a plastic gasket to insulate between the door and the frame.

Storm Doors

Many homeowners add a storm door to their exterior doors. This is a lighter door hung in the same frame with the exterior door. They are made of either wood or metal (usually aluminum) and have an interchangeable window panel or screen panel to convert them to a screen door in the summer. Some aluminum doors have a window which can be slid out of the way of the screen for summertime use.

A storm door cuts down on the amount of cold transmitted through the main door. The 3 to 5 inches of space between the doors provides some insulation, but since both doors are open at the same time when someone is entering or leaving, they don't stop the blast of cold air when they are opened.

Caulking Needed

Window and door frames installed in a new house should have non-hardening caulking material on the under side of the brick mold or trim as they are put in the opening, so there will be no passage for cold air if there are irregularities in the house sub-siding material. You can use painter's caulk, with latex and silicone, or straight silicone. Other materials that can be used are thin plastic foam or rubber weather stripping material that will compress into any small crevices. Putting in caulking or weather stripping material as the window is put in the opening is preferable to caulking around the frame after the window has been put in place.

Caulking the outside of a window or door frame is only part of the job. The space between the door or window frame and the frame of the rough opening should be filled with insulation. Just applying caulking to both sides of the frames doesn't do the whole job.

When caulking around windows and doors in older houses it may be necessary to remove the inside trim and fill the space between the frame and the rough frame with insulation. This and a bead of painter's caulk along the edge of the outside trim and the house siding will drastically reduce air movement around windows and doors. Painter's caulk is made in several colors, won't harden, and can be painted. It is usually an acrylic latex or acrylic latex with silicone material and is available in small tubes that fit into a caulking gun. Pure silicone caulking may also be used, but it is more expensive and more difficult to paint.

Installing a Window

Some double-glazed windows may be installed without a regular frame around them. If you put in one of these or replace one that has been damaged, you must observe some precautions. The window should be approximately ½-inch smaller than the opening, to allow for at least a ¼-inch space on all sides. The window should be placed on either hard rubber or soft wood blocks, which are obtainable from most glass companies that replace these windows.

The reason for the space all around the glass is to allow for any movement in the house frame or expansion of the glass from the heat of the sun.

The blocks are the exact thickness of the window, and about 4 inches long. They should support the window in only two places, so there will be no chance of any distortion in the frame causing a strain on the window.

Caulking

Caulking should be acrylic latex with silicone unless the company furnishing the window recommends some other type. Do not use straight silicone as it will stick to the window tight enough to crack it if the wall shifts. Caulking should be put in only two places, between the framing stops, the thin boards that form the track the windows fit in, and the window, and the stops and the window frame. Do not fill the space between the window and the frame with caulking as this will not allow any movement of the house without stressing the window.

Doors and windows will come from the factory with weather stripping in place. In most cases this is easily replacable with material from the original supplier, by following the manufacturer's directions. The material may be either a plastic or metal strip that fits into a prepared groove in the window or the frame, and seals it by pressure caused by its shape. It is the most effective weather strip, but will wear out in time.

For an older house that has never had any weather stripping, it may be a job to put in the friction type material. You will have to remove the windows from their frames by removing the interior stops. In an old house, figure on replacing these, as they will probably be damaged when they are removed.

Grooves will have to be made (usually a saw cut) in the window frame to accept the weather strip unless it is designed to be nailed or stapled in place. When the weather stripping is in place, the window is put back into the frame, and the stop renailed back against it.

Other Alternatives

For a house where it would be difficult to remove the windows, and for doors that can't be done with the friction-type weather stripping, there are several materials that can be nailed or stapled on. These come in rolls or rigid strips which can be cut to length and fitted along the window or door frame on the outside with the window or door closed. Apply just enough pressure to the strip to seal any cracks along the frame, and then nail or staple the strip in place. These weather strips are good for one or more seasons, depending on how much the window or door is used. As these strips are plastic foam, rubber, felt, or one of the various plastics now available, and backed by wood, plastic, or metal, they can be easily damaged. If they get wet, they may freeze to the door or window and be destroyed when it is opened. However, they are much better than the rags or newspapers used to seal cracks in years gone by.

Homeowners should take pains to get the best doors and windows available, taking into consideration their budgetary restrictions. The owner of an older house can do a lot by caulking and weather stripping the doors and windows. If the windows are showing signs of dry rot, they should be replaced with newer, more energy-efficient units. These will, over time and like insulation, pay for themselves.

An added note of caution here. Many window manufacturers are making windows with a tinted glass that they claim has a higher R-value than regular glass. It is also supposed to reduce heat from direct sunlight. While better building materials are being produced all the time, as with everything else, some of the claims are slightly exaggerated, or may be the results of controlled tests.

Homeowners should weigh the supposed benefits against the extra cost and decide just what they want. A check with your local university Extension Service or an organization such as NATAS (address in the appendix) should provide information either proving or disproving any claims about what these products will do.

18. Earth Houses

A CONSTRUCTION TECHNIQUE that picked up a considerable following in the 1970s is the earth house. This is a house buried in the ground, either partially or completely. Most designs use partial exposure. One reason this type of construction became popular was the claim the temperature of the earth, at depths of 6 feet or deeper, stays at a constant temperature of 55°F. That depends on where you live.

Some areas of the contiguous states have wintertime frost levels as deep as 12 feet, depending on the snow cover. In Alaska and northern Canada, south of the permafrost line, frost (the freezing level) may go as deep as 20 feet. The temperature at the frost line is 32°F. Above that it is colder. Below, warmer. To pour concrete in the ground and expect it to remain at 55°F is a fallacy.

Even if it did, without some insulation between it and the surrounding soil, the concrete would remain at 55°, or 15° colder than most people find comfortable. The best move is to keep the cold away from the concrete, not try to warm the room inside the concrete and hope the concrete will get that warm. Again, concrete conducts cold. If it is a given temperature on the outside, at best it will only be a few degrees warmer on the inside.

Use Rigid Insulation

As with a basement, it is easier to prevent the cold from entering the concrete than to try to warm it. Polystyrene, polyurethane, or isocyanurate rigid insulation can be used to insulate straight walls on an earth house, the same way they can be used to insulate a basement. The concrete needs to be sealed to prevent moisture intrusion, and the insulation protected with a layer of polyethylene sheeting between the insulation and the soil surrounding the building.

If the earth house is a more modernistic design, with a dome or with rounded corners, a spray-on insulation such as polyurethane foam can be used. A layer of uniform thickness can be placed onto virtually any shape house. Depending on the material used, foamed insulation may need some form of protective covering. If nothing else, it should be protected by a layer of polyethylene sheeting between it and the soil. For the same reasons that a basement floor benefits from insulation, the floor of an earth house should have it. You don't want floors that are 15° colder than the rest of the house.

Store Heat

Insulating a concrete wall from the cold also allows the homeowner to take advantage of stored heat. Insulated concrete absorbs heat, up to the temperature of the room it surrounds. Once at that temperature, it takes less energy to maintain a given temperature. Insulation is what guarantees a constant temperature inside an earth house, not the temperature of the soil surrounding it.

19. Brick, Block, Stone, and Concrete Houses

FOR CENTURIES, until the advent of balloon framing, virtually all houses were built of stone. Even those buildings using a wood frame such as post and beam often had stone as the filler between the posts and beams. In the New World, where there was an abundance of stone, the immigrants kept the old-world tradition and used stone for the more substantial buildings. It was solid, it was permanent — and it was cold.

Many of the old stone houses are still in use, and modern builders build from stone-like materials such as brick, concrete blocks, and concrete. Houses built of these materials are still solid, permanent, and, unfortunately for the modern owner who is concerned about the cost of energy, cold in the winter and hot in the summer.

Several methods of insulating these houses have been tried over the past several years. Houses were built of concrete or other similar hollow blocks (pumice and cinder) that had expanded mica poured into the hollows in an effort to provide some insulation. A quick look at the chart of R-values shows the results. With concrete webbing reaching all the way through the wall, the small pockets of insulation did very little to stop the cold from penetrating.

Insulation Uses

It has become apparent to many owners of "stone" houses that some major construction was needed to provide enough insulation to make an energy-efficient house. This most often takes the form of building a house within a house by building a frame wall at the proper distance inside the stone wall to allow installation of enough insulation to make the house energy efficient.

In most areas it also entails putting in a vapor barrier, the same as for a frame house. This limits the amount of moisture that can condense on the inside of the concrete or brick walls. In addition, for a block or brick house, it limits the damage moisture does to the mortar between the blocks and bricks. Because of its chemical composition, mortar is degraded by moisture escaping from the building, and will begin to crumble and fall out after several years. When this happens, it must be replaced.

Repointing Expensive

In our nation's capital this replacing or repointing of the mortar in our public buildings costs many thousands of dollars every year. The same is true in many college buildings in the northern states. Without a vapor barrier, this is an ongoing expense for the owner of block, and especially brick, buildings.

Some brick houses built in the 1950s and 1960s present another problem. They were constructed with a frame wall inside the brick walls, so the interior finish (usually gypsum board) could be more easily installed.

This also provided a place to put the electrical wiring in the outside walls. Unfortunately, the space between the bricks and the finished wall was usually 2 inches or even less. This isn't enough space for adequate insulation, even if there were some way to get it into the wall. Foamed insulation could be put in most of it, but 2 inches of that still isn't adequate in most areas of the country. The most cost-effective solution is to tear out the original false wall and put it back far enough from the bricks to give the necessary depth for insulation.

Brick Veneer

Brick veneer houses have thin bricks fastened to a layer of mortar held onto the plywood or other outside sheathing of a standard framed wall. In newer houses these framed walls are usually insulated. Depending on the climate, they may not have enough insulation for true energy efficiency. If this is the case, the homeowner can use one of the methods discussed in the chapter on insulating walls in older homes.

20. Log Houses

It may come as a surprise to some, especially owners, but many log houses should be insulated to make them energy efficient. A look at the chart of R-values will tell one reason why. Dry wood can have an R-value of up to 1.25 per inch. Low quality insulation has a value double that and more.

Another problem not often considered is moisture. Where does the moisture in the house go when the outside wall is solid wood? It still tries to get out, and might, if there are cracks in the logs.

The propensity for moisture-laden air to move outside has been used in the past to make a log house tighter. In the far north, where winter temperatures may stay below freezing for months, it was a common practice to seal the house from the inside. This was done by putting several containers of water on the stove and bringing them to a boil. The vapor traveled to the outside through the cracks. It immediately froze on contact with the outside air, effectively sealing the house until the outside air temperature again was above 32°F. This cut down on drafts and made for a snug house, at least until the weather warmed up.

Logs have another property that helps explain why log houses don't seem to be as cold as the R-value of the logs would suggest. They are capable of storing heat. Logs in a house wall will absorb heat up to the temperature of the air in the house. When the house temperature is lowered, say for the night, they give back heat into the house until the temperature is again equalized. Unlike a block or stone wall, logs can be at two temperatures at once, one on either side.

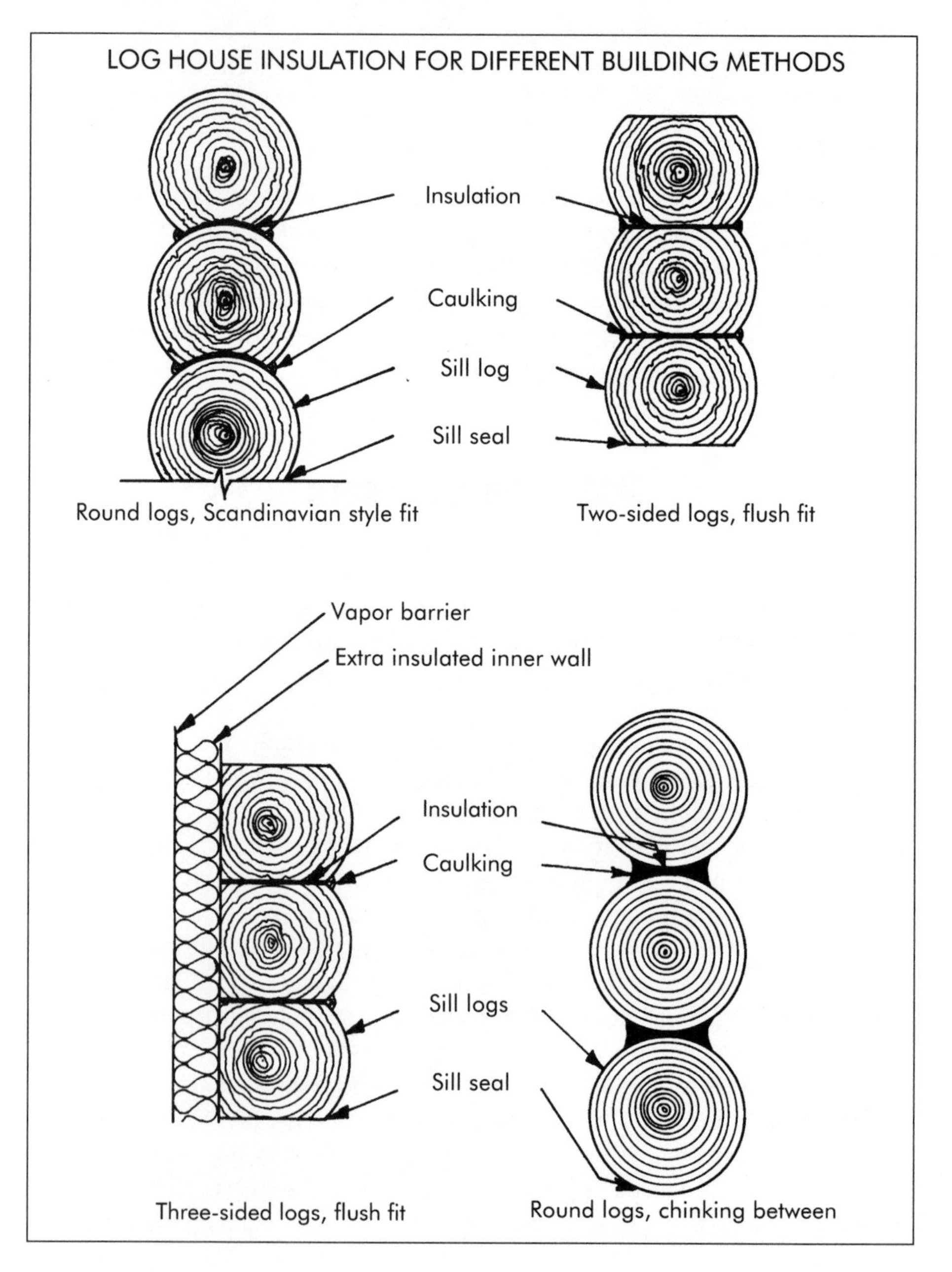

Chinking

Basic insulation for a log house, even if none is required elsewhere, is needed between the logs. No matter what type of log construction is used, round fitted, flattened, three-sided, or round with chinking, a thin layer of insulation should be put between each log and those above and below it. This fills irregularities between the logs and closes cracks open to the outside. Caulking or "chinking" is then done on either side of the insulation.

The size of the logs makes a difference. Many modern log houses are made of logs 8 inches or less in diameter. Two-sided and three-sided logs are often only 6 inches thick. When placed in a wall, there may only be 3 or 4 inches of contact surface between the logs. This contact is the thickness that should be considered when thinking of additional insulation. Not only is the joint between logs thinner than the logs, but that is the place where there will be leaks in the wall if there are any.

Even houses made of turned logs, which are logs turned to a given diameter in a large lathe, then grooved with a machine to give a very tight fit, need insulation and caulking where logs meet. This type of construction is a machine-made variation of the old Scandinavian system of building log houses. Most of these houses use logs turned to 20 inches or less in diameter, with only 8 inches or so of thickness where the logs rest on each other.

Some log house builders still hand-fit the logs in the same Scandinavian style. They usually use logs from 20 to 36 inches in diameter, and will have a thickness where they join of 12 inches or more. This type of construction is usually done only on contract.

Factory-built log houses are most common today. The logs may be only 6 or 8 inches thick, and often are squared on two or more sides for ease in construction. One maker of log home kits even starts with 8x8 timbers, running them through a machine that rounds one side, which is for the outside of the cabin, and puts a tongue-and-groove in the top and bottom for a tight fit. The inside surface of the log is dressed smooth so it looks like lumber. This construction needs a strip of insulation between the "logs," and caulking in the crack between the logs on both the inside and the outside.

Peeled Logs

Some log home builders use hand-peeled logs, without doing anything else to them. The bark of the log is removed by a man wielding a large drawknife, then the logs are fitted into the house by notching the ends so there will be as small a crack as possible between each set of logs. Usually the logs are no less than 8 inches in diameter on the small end, and no larger than 14 inches in diameter on the large end. As a wall is built, the large end of one log is placed atop the small end of another to get a close fit between the logs. Once the logs have been fitted for a house at the factory, they are numbered and trucked to the home site, where they are reassembled.

Once the house is up, the gaps between the logs are stuffed with fiberglass insulation, then a vinyl material that won't shrink away from the logs is applied, sealing the crack. The wall between the logs may be only 3 inches thick, and is filled with a material that has a lower R-value than the logs themselves, so the insulation value is low.

Roofs

Log houses often differ from frame houses in the roof. In order to get the most of the rustic feeling from a log house, the roof is often open beam. In that case the insulation has to be put on top. In the good old days, sod was often the material used, with up to a couple of feet being placed on top of the house. This entailed a heavy roof structure as sod can weight up to 75 pounds a cubic foot. Even though sod has approximately the same R-value as fiberglass, nowadays it will probably be more expensive.

What we now might term as rustic or old-fashioned was, more often than not, done out of sheer necessity by our forebearers. Log houses were built because there were more than enough trees, and the trees needed to be removed to clear land for growing food. Sod was also plentiful, especially in places where there was some prairie. It was there for the taking. A couple of men with a saw and axes and a flat shovel, often carved from a split log, could build a solid house in a couple of days, with the expenditure of nothing more than sweat.

They also weren't concerned about fuel to keep the place warm. If it was cold, they built the fire up more. Energy-efficient wasn't in the vocabulary. Insulation also wasn't in the vocabulary, except in the far north country where tundra moss was used to stuff the cracks between logs or stones, and often put in a thick layer on the roof.

During experiments on the insulation value of indigenous building materials, University of Alaska engineers discovered that dried tundra moss has a higher insulation value than most manufactured insulations. Of course, unless you live on the tundra, the cost of obtaining the moss puts it out of reach. It is also very flammable and would probably be banned by most fire codes.

Adding Insulation

Most log houses only use insulation between the logs. However, you must do something to increase the insulation along the logs to make yours a truly energy-efficient house. This means building a wall on the inside of the log wall. (Note diagram.) This should be thick enough to provide energy-efficiency. It should also have a vapor barrier on the warm (room) side to reduce the moisture flow into the log wall as much as is possible. Once the vapor barrier is in place, the wall can be finished as desired.

This is extra expense, but if the log walls of the house are as thin as some described earlier, it may be necessary for the conservation of energy (and money).

If homeowners don't want a sod roof on their log houses, they can build a roof as described in the chapter on insulating the roof of a new house. They can also use the insulated panels described there. Vapor barriers and ventilation are still a consideration. A roof on a log house has the same problems as one on a frame house, and should be treated the same way.

In order to get the rustic effect, and utilize the efficiency of modern insulation, some log home owners have had their roof foamed with urethane foam insulation, in some cases then adding a layer of sod to complete the illusion of a rustic dwelling. The same problems and benefits apply here as they do for a regular framed roof. The point is to get the highest insulation value for the money spent.

21. Metal Buildings

WHEN MOST PEOPLE THINK of metal buildings, they think of the buildings used on farms for machinery and product storage. These usually aren't insulated, except for some part used as a workshop which may be heated. Even when insulation is installed, it is often inadequate because the heat is only needed while people are actually working there. A large, inefficient wood or oil-fired stove is often used.

At this time, companies are manufacturing metal buildings for use as homes, churches, and virtually any other use that is wanted. These are prefabricated buildings. The parts are made in a factory, then hauled to the building site and quickly assembled. Prefabricating and quick assembly are the main selling points of metal buildings, as labor is the most expensive part of any building.

Metal, by its very nature, does not insulate, either from hot or cold. It is a very good conductor of both. A metal building without insulation will be hot in the summer and cold in the winter. The attic of a metal building will get extremely hot in the summer and as cold as the outside area in the winter.

Ventilation is more important in a metal building than in a wooden one, as metal is an effective vapor barrier. Thus, any moisture in the building will condense on the inside of the exterior sheathing if the outside air is below the dew point of the inside air. It follows that insulation placed in the walls and roof of a metal building should be kept away from the siding and the roof metal, to prevent it from soaking up moisture from them. An inch of space between the insulation and the metal will also allow for air circulation, and with a good vapor barrier on the heated side of the wall, will take care of most of the moisture problems.

Problem Ignored

The builders of early metal building kits didn't address the problem, so in many the insulation became water-soaked and often fell off the walls and ceiling. The reason for this was the way the insulation was installed. Most buildings were shipped with R-11, or 3½-inch thick fiberglass insulation. The building frame was assembled, then the insulation was put on the outside of the frame, then the siding/roofing over that. The sheets of metal were held to the metal frame by "self-drilling" screws which went through the insulation between the metal sheets and the frame members, compressing it to about ¼ inch thick. The insulation was backed with either aluminum foil or vinyl, and was held tightly against the outside sheathing by the building frame, so it soaked up any condensation on the metal walls.

Users also found that when the outside air was below freezing, moisture in the inside air became frost on the screw points inside the building. In the roof, these would sometimes become icicles several inches long. When it became warm enough for them to melt, they rained onto whatever or whoever was in the building.

When urethane foam insulations became available, many users of steel buildings believed they had found the solution to their problems. The urethane was sprayed on the inner surfaces of the building, and moisture problems were (almost) solved. Unless the whole steel frame was covered with insulation, there was still the problem of transmission of cold through the steel.

One problem was solved; another presented itself. Because urethane was flammable, building codes required it be covered by nonflammable material, generally ⅝-inch thick gypsum board. Since most of the framing of a steel building was on wider spacing than was required for gypsum board, something else had to be added to support the gypsum board. This was either wood or light steel studding or furring strips spaced 24 inches apart or closer, depending on code requirements.

The Solution

What has evolved is a building within a building, in order to properly insulate steel buildings. A wood or light steel framework must be constructed inside the main steel building frame, and insulation placed in all available space, especially between all building frame members and the inside sheathing. A vapor barrier is then added, and the inside sheathing put in place, as with any building.

Unfortunately, what was believed to be an inexpensive building rose in cost until it was cheaper to build from wood and insulate in the regular way. Steel buildings are less costly than any others if there is no need for insulation. However, to install insulation can easily raise the price to where another form of construction can be cheaper, both in the initial purchase price and ongoing cost of energy to heat or cool the building.

Other Insulation Techniques

22. Insulating Heating Ducts

WHILE NOT A COMMON PRACTICE, insulating heating and ventilating ducts is another of those little things a homeowner can do to increase the distance his energy dollar will go.

Forced air heating systems and air-conditioning systems move air throughout a house by means of metal or plastic cloth pipes, commonly called ducts. A modern house with a forced air heating system (which also doubles as an air-conditioning system during warm months) can have as much as 200 feet of ducts ranging in size from 12 x 30-inch metal boxes near the furnace/air conditioner to 5-inch diameter pipes near the outlets.

If any of these ducts are in an unheated (uninsulated) crawl space or above the insulation in an attic, the air in them will quickly be at the temperature of the surrounding air when the system fan is not operating. This means that when the system fan starts, there will be a blast of very cold (or warm) air before the system goes to work. Also, the heat or cold in the ducts will radiate into the air surrounding them, using extra energy to heat or cool space not used for the occupants of the house.

The heating and ventilating industry has designed several ways to overcome this problem. Special duct insulation, which is 1 inch thick rigid fiberglass with aluminum foil backing on one side, is available in most building supply and heating supply firms.

How to Install

Two items are available to install this insulation. Stickpins are nails about 1⅜ inches long, with very sharp points and 1 inch square heads with a powerful "peel and stick" adhesive. They are used to hold the insulation in place. An aluminum foil tape is used to cover corners and joints. This is designed only for rectangular ducts. Regular backed fiberglass roll or batt insulation can be used for ducts of other shapes.

You can avoid having to insulate round metal ducts by using insulated flex ducts. These are plastic tubes with a spiral wire imbedded in them to help them to retain their round shape. They are covered with an inch of fiberglass insulation and a plastic cloth cover. Flex duct, as the name implies, is very flexible and can be routed through places where it is difficult to put metal tubing. Insulated flex duct is frequently used in attics and other places where space is limited.

Uninsulated flex duct is also available. Since it is more expensive than metal pipe, use it only where odd bends must be made, or places where it must be pulled through floor joist spaces, such as in the floor of a two-story house being converted to a forced air heating system.

Using Fiberglass

Installing rigid duct insulation is simple. The first step is to cut the insulation to size, since it is sold in 4 x 8-foot sheets. The long edges are formed to make a lap joint, so cut it across the 4-foot dimension instead of lengthwise.

To cut rigid fiberglass, lay it on a solid surface, such as a piece of plywood, and cut from the foil-covered side, using a utility knife. A Sheetrock square is helpful in making straight cuts.

Cut one dimension 2 inches wider than the duct, to cover the corners completely. For example, if the duct is 8 by 24 inches, cut two pieces 24 inches wide and two pieces 10 inches wide.

Make sure the duct is clean and dry. It should be at room temperature. Peel the backing off the heads of the stickpins and press them firmly onto the duct surface, two rows about 12 inches apart on the 24-inch sides of the duct, and a single row in the middle of the 8-inch side. Use four pins for each of the larger pieces of insulation, and two pins for the narrow pieces.

Align the pieces of insulation that are the exact width of the duct with the duct edges and, with the aluminum foil side up, press them firmly down onto the stickpins. Be careful where you put your hands. Those pins can make a nasty wound.

When the pins come through the foil, use the small square pieces of metal that come with the stickpins to hold the insulation in place. The retainer is pushed down onto the stickpin, with the small indentation away from the insulation. Do not use excessive force. The retainers should be only tight against the foil, not compressing the insulation.

Put the other two pieces on, making sure the cut edges are aligned with the other two sides so there is a smooth corner completely covered with insulation. Cover the entire duct from end to end, using the lap joints of the insulation to make tight joints between each section of insulation.

Tape the joints first with the aluminum tape. Then tape the corners, making sure the tape reaches from the aluminum backing on one piece of insulation to the backing on the other piece. Go back over the tape seams, pressing them firmly onto the backing on the insulation.

As a safety precaution, once the retainers have been put in place, cut the sharp points off the pins.

Insulating Round Ducts

To insulate round ducts with regular batt or roll insulation, lay the insulation lengthwise on the pipe and bring the edges together, holding them in place with regular duct tape. Avoid compressing the insulation. The backing on the insulation should be on the outside, away from the pipe. Use insulation wide enough so the edges will easily meet. The paper strips along the edges of the insulation may be stapled together for added security, using a plier type stapler or a small desk stapler.

Insulating Against Noise

Two other types of insulation are used with heating/cooling ducts. Both are used to insulate against noise. One is a rigid material 1 inch thick and with no backing. It is used primarily where there is a possibility of noise from the movement of air in a duct, most often behind return air grills. It is held in place by the same stickpins used for the rigid outside insulation. It may also be glued in place, using a glue made for that purpose. It must be fastened securely. If it comes loose, it can be pulled into the duct by the air movement and plug the duct or get into the furnace air chamber, disrupting the air flow through the system.

Another form of noise insulation is a strip of material about 5 inches wide, with a metal strip on each edge separated by a flexible fiber strip. This is placed between sections of metal duct to eliminate the transmission of vibration along the duct to the heat vents in the house. It is most often placed near the furnace, sometimes where the heat ducts link with the furnace body.

Thicker insulation may be put on rectangular ducts by using either polystyrene or thicker rigid fiberglass. Rigid fiberglass is available up to 4 inches thick. This material must be glued on, using a glue that will bond it to metal. If you use unbacked foam insulation, you do not need to add backing. Don't force foam insulation tightly between framing members of a house, such as the floor joists, and ducts. Any vibration of the duct can cause a squeak where the insulation contacts the wood.

Plan in Advance

Insulating heating/cooling ducts is something else that should be planned. This is especially true for ducts which traverse an attic that will be insulated. If the ducts are installed before the attic is insulated, they may be placed so the insulation blown into the attic will cover them, too. This will eliminate the need for special duct insulating.

Often ducts are placed in soffits in a room, along an outside wall. The duct is put up before the wall is finished, and the finish material is placed around the duct, to conceal it. Insulation should be put on the outside wall before the duct is placed, to prevent any cold spots along the duct.

Similarly, a duct beside a foundation wall in a crawl space needs to have insulation between it and the foundation to avoid cold being transmitted to the duct from the outside through the foundation wall.

In short, any place where a heating or cooling duct may be affected by heat or cold should be insulated. A lot of heating/cooling contractors consider insulating duct work to be a nuisance, but for a homeowner concerned about getting the most for his energy dollar, it is something that should be done.

23. Insulating Water Heaters and Pipes

THERE ARE SEVERAL REASONS for insulating a water heater and the pipes that lead to and from it. A primary reason is to save energy dollars. Depending on the climate, this may not be the main reason for plumbing insulation, since this also contributes to the overall comfort and convenience for those living there.

In the United States, most water heaters are operating all the time. Water is kept at a given temperature 24 hours a day. Whenever the temperature of the water in the heater falls below a pre-set point, the heating element turns on, raises the water temperature back to the pre-set point, then shuts off.

Use Timer

If no one is in the house, as is the case in many homes during the day, this constant reheating is wasteful. One method for dealing with this is to put a timer on electric water heaters which causes them to operate only during times of peak demand. This saves 10 to 25 percent of the electricity used to heat water in a standard home.

It is easy to install a timer on an electric water heater. It isn't easy to install one on a gas-fired unit. Therefore some other method of regulating the amount of time the heater operates is needed. This is provided by adding insulation around the body of the heater, thus reducing the heat loss from the unit and reducing the number of times the heating element must operate to maintain the pre-set temperature.

In times past, many water heaters had no insulation. When the water was hot, the tanks radiated heat into the surrounding air at a high rate. Manufacturers began putting a jacket of light metal around the tank to protect the users, as it was easy to get burned on an exposed water heater. They also wrapped the tank in a thin layer of fiberglass, probably as much to keep the outer jacket from contacting the tank as to keep heat in.

Heaters now are being made with foam insulation, but it is still only ¾ to 1 inch thick, so has a low R-value. On virtually all water heaters, the heat from the water can be felt on the outside jacket. This has led many to conclude that another layer of insulation placed outside the water heater's jacket will result in saving energy. Some sources say this can be as much as 25 percent of the cost of heating water.

Blankets Available

Water heater blankets are available from building supply houses, plumbing suppliers, and many other sources, at a variety of prices. Most of these are an inch or two of fiberglass bonded to a vinyl cloth backing. Some are made to fit specific sizes of water heaters. Others can be cut by the installer for any size unit. They can be used on both electric or gas-fired heaters, with some precautions on the latter.

Some blankets are designed to cover the entire heater, including the top. Those for gas heaters are a simple wrap for the main part of the heater, as safety demands that nothing blocks the flue or the burner areas, to eliminate fire danger. The R-value of these blankets is low.

Many homeowners make their own water heater blankets of fiberglass insulation batts, 3½ to 5½ inches thick and held in place with duct tape or something similar. For an electric water heater, this is inexpensive. The blanket can also be used on a gas water heater, but allowances must be made for the opening to the burner, and it cannot be put on top of the heater, to avoid blocking the draft hood.

Insulating Water Pipes

Insulating the water heater is only part of the job, though most homeowners stop there. The pipes leading from the heater to the faucets should also be insulated. There are two reasons for this. First, the copper pipe that is most often used is a heat conductor and radiates the heat from the water in it into the surrounding air very quickly. (Hot water heating units use copper pipe because of this ability to radiate heat efficiently.)

The other reason is to reduce heat loss in the water, so less is used each time a hot water tap is turned on. A water heater should be located so the water travels the shortest distance possible from the heater to where it is used. This requires planning.

There are several ways to insulate water pipes. One is to use the foam pipe insulation made to fit each size of pipe. This puts approximately ½-inch of insulation around the pipe. It is made in the form of a tube, and is split on one side so it can be slipped over a pipe. Some brands have the edges of the slit coated with a glue, protected by a plastic strip so it doesn't stick before the user wants it to. Once the insulation is in place, the plastic strip is pulled off and the gummed edges of the slit are pressed together, making a solid wrap around the pipe. The plastic tubing can be formed around corners, and cut to any length needed, using a utility knife or heavy scissors.

Fiberglass Cover

A less expensive pipe insulation made of fiberglass is also available. This comes in widths of 2 to 4 inches. Some brands have the fiberglass backed with vinyl tape, and others have the tape separate, so it must be put on after the fiberglass is in place. This strip insulation is wrapped around the pipe in a spiral, the same way the handle of a baseball bat is taped. The job must be done carefully so the fiberglass insulation isn't crushed and its R-value drastically reduced.

If pipe insulation is part of a larger insulating job where strips of backed insulation are left over, it can be used instead of the special pipe wrap. It may be either wrapped in a spiral, as with the pipe wrap, or placed lengthwise on the pipe, wrapped around it, and held in place with duct tape or something similar that won't dry out and come loose over time.

Pieces of rigid insulation can also be used. Cut a groove in the insulation, so that the pipe fits between two pieces of the material without bending the insulation. Hold it in place with tape, as with the strips of fiberglass. These methods use pieces of insulation available from another job. Otherwise it is easier to use insulation designed especially for pipes.

Two Reasons

There are two other reasons for insulating plumbing. One is to eliminate problems from condensation on cold water pipes; the other is to prevent the pipes from freezing.

Where plumbing is in a crawl space, condensation doesn't seem to be a problem to the homeowners. That is because they never see it. When the temperature of the water in a pipe is below the dew point of the surrounding air, there can be condensation on the pipe. The higher the humidity, the greater the amount of condensation.

If a homeowner stores things in the crawl space, the condensation will drip onto and may damage the stored items. Condensation on a pipe can be heavy enough to convince the unknowing that the pipe is leaking. All types of pipe that can be used for cold water lines are subject to consideration if the conditions are right.

The way to eliminate this condensation is to insulate the pipe. This is done with the same materials and in the same way that hot water pipes are insulated. However, the job must be done when the pipe is completely dry. Otherwise the insulation will get wet and not do the job. The pipes must be insulated all the way to where they enter the building, to prevent any part of them from causing condensation which may drain into the insulated part of the pipe and ruin it. In climates subject to high humidity, this may be from where the water supply line enters the building to where each separate cold water line goes through the floor to whatever plumbing fixture it supplies.

Toilet Condensation

Condensation is also a problem with toilets in some areas. Even the short supply line from the wall to the toilet will show some condensation, especially if the humidity in the house is higher than that outside. Also, if the toilet tank isn't insulated, moisture will condense on it and drip onto the bathroom floor.

There are three ways to deal with toilet condensation. One is to install a mixing valve in the water system for each toilet in the house. This mixes warm water from the hot water system with the cold water going into the toilet tank, bringing it to room temperature, so the moisture in the room air won't condense on the toilet tank.

Interior Insulation

Another method is to have insulation *inside* the toilet tank. Many newer toilets come from the factory with insulation already installed.

Insulation kits may be purchased from building or plumbing supply firms. These are made of several sheets of polystyrene insulation ⅜ inch thick, with glue to hold them in the toilet tank.

To install an insulation kit, drain the tank and wipe it dry. Cut the sheets of insulation to fit, starting with the bottom, apply the glue to them, and put them in place. The directions with the kit will tell how long to wait before refilling the tank. With a little care during installation, this insulation will last a long time.

The last method, usable in areas where condensation is a problem only part of the time, is to wrap the outside of the toilet tank. Since most homemakers would object to the looks of something made of fiberglass insulation, home decorating suppliers now offer a tank sleeve made of the same material as seat lid covers. If the plumber left a half-inch or more of space between the tank and the bathroom wall, these sleeves can be slipped around the tank, effectively insulating them from the room air and eliminating condensation problems. The decor can also be changed to fit the mood of the decorator at any time. A drawback is that the insulating material must be removed and washed periodically.

Danger of Freezing

Both water and sewer pipes should be insulated if there is danger of them freezing. Even that "once in ten years" cold snap can do a lot of damage to water lines and, as a result, to your bank account. In climates that have much freezing weather it is strongly recommended the crawl space be insulated and the vents be closed during the winter. Heat should also be provided in the crawl space to prevent the pipes from freezing and to eliminate cold floors.

Where a house is built on pilings, or a climate where it may be necessary to keep the crawl space vents open all the time, the plumbing should be insulated to safeguard it from freezing during that "once-a-winter cold snap." In warmer climates, the insulation used to prevent condensation will also prevent freezing. However, if there may be several nights of freezing weather during the winter, heavier insulation may be needed, or some way to warm the pipe enough to keep it from freezing.

Heat Tape

A heat source is often a device called a heat tape, an electrical unit with wires that are laid lengthwise of the pipe, or wrapped around it in a wide spiral. Most have a thin layer of insulation wrapped around the pipe over them, to hold them in place and concentrate the heat against the pipe. Heat tapes are not designed for constant use. If operated that way they may burn out, leaving the pipe unprotected. To prevent constant operation they generally are protected by a sensor or thermostat that only turns them on when the air or pipe temperature drops to 35-40°F. It would be prohibitively expensive to heat tape a lot of plumbing. Insulating the crawl space is still the best way to go.

For a mobile home or a house built on pilings, where the only exposed plumbing is from the ground to the insulated underside of the building, wrapping the pipes with a heavy layer of insulation may suffice. However, as these dwellings often have the entire area under them open to any type of intrusion, animals may damage the insulation. If it is not repaired, the pipes may freeze.

Utilidor

This is where a device called a utilidor comes into use. This is an insulated box, usually of wood, and the protected pipe runs through the center of it. The box may be made large enough to contain enough insulation to protect the pipes in any weather, or it may contain just enough insulation for a normal winter, and require a heat tape againsts the pipe, just in case.

In a climate where the ground freezes to any depth, extend the utilidor into the ground to or beyond the normal frost depth to prevent pipes from freezing below where they are protected.

While taking care of the problem of freezing, this brings up another one; that of the ground moving up or down from the frost. It is best to build a "sliding joint" utilidor when both ends are against something solid such as the ground and the underside of the dwelling. Water and sewer lines inside the utilidor should have a flexible section to prevent damage to them if the ground moves. (See illustration.)

Utilidors may also be used for water and sewer lines laid horizontally, such as lines going to an outside building. (Check your local building code to find out if water and sewer lines can be put in the same utilidor.) They can be buried, laid on top of the ground, or suspended in the air. They may be made solid, with enough insulation to keep the pipes from freezing, or hollow, with insulation several inches away from the pipes and some means of forcing warm air through the utilidor. A solid utilidor can end against the side of a building or anywhere that is solid. A hollow utilidor must necessarily have both ends at some place where an air flow won't be restricted. As with other insulatiion methods, the use of utilidors is restricted only by the ingenuity of the builder.

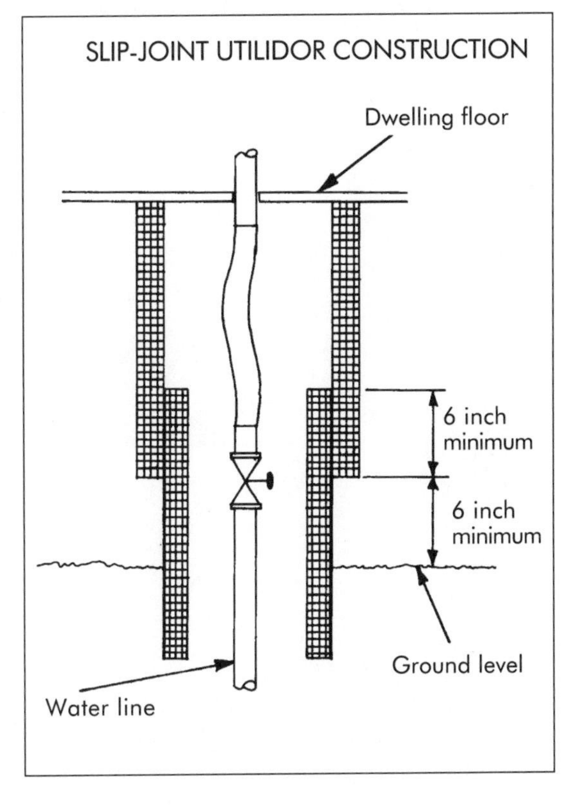

Heated Area

Another method of insulating unprotected pipes from the ground to the underside of a building that is similar to a utilidor is a small heated room around the area of the plumbing. This may be several feet square. It is well insulted, with about the same amount of insulation as is used in the walls of the building.

A source of heat, such as a warm air duct, is put in. This system partially relies on the building's heating system for protection of the plumbing. It has the same problems with the same solutions as a utilidor where the ground may freeze and move under it.

Underground Protection

In many parts of the U.S. and Canada, water and sewer lines can be laid directly in the ground at a depth that is normally below frost level. Again, the exceptionally cold winter may cause them to freeze. An inexpensive protection for such lines can be provided when they are laid. Put the line into the ditch in the normal way, and put down a layer of rock-free dirt or

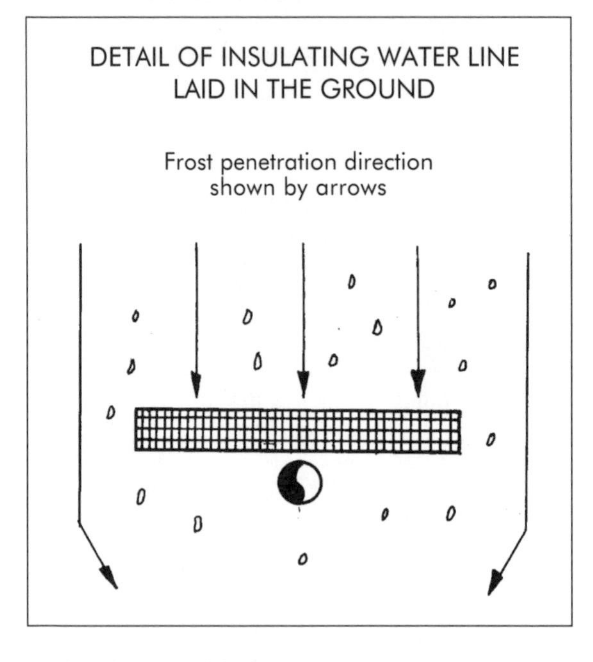

sand to cover it 2-4 inches deep. Level this and put a layer of 2-inch polystyrene insulation 24 inches or wider on that, with the pipe centered under it. (See diagram.) This prevents the frost from penetrating the ground around the pipe, keeping it from freezing. Even if a pipe is buried well below any danger from frost, putting a layer of insulation over it where it passes under a driveway or road will prevent traffic from driving the frost down to where the pipe could freeze.

Another pipe insulation method used by some ingenious individuals is to place a water pipe inside a larger plastic pipe, such as that normally used for a sewer line. The water pipe is positioned directly in the center of the larger pipe, then foam insulation is forced into the larger pipe completely surrounding the water line. This might be a solution in places where the water line cannot be buried deeply, and there is only a moderate danger of freezing.

24. Insulating for Cooling

INSULATION ISN'T just to keep the heat in during cold weather. It also keeps the heat *out* during hot weather.

In chapter two, on how to figure the insulation needed, the formula uses heating degree days as the basis for the calculations. It was mentioned briefly that heating degree days and cooling degree days could be combined in climatic areas where they are nearly equal in number.

Cooling degree days are figured the same way as heating degree days, except that days with temperatures *over* 70°F are used as the basis for calculations. The number of cooling degree days for a given locality can be obtained from the same places you would find out heating degree days: The local university Extension office, the National Weather Service, or your local electric utility company.

Once you know the cooling degree days for your climate, (or the combined heating and cooling degree days), the formula is the same as given in chapter two. Multiply the total degree days by .004. The answer will be the R-value of the attic insulation needed. The percentage numbers for the walls and floor are the same as given previously.

Floor insulation should be used where there is a well-ventilated crawl space or the house is built on pilings. If a house has a basement, it may be necessary to insulate the part of the basement wall that extends above ground level, as concrete is as good a conductor of heat as it is cold. Outside insulation (to keep the heat from the concrete) is best, but insulation may also be put on the inside, the same as in a cold climate.

Vapor Barrier

The major difference between insulating for cold and insulating for heat is the vapor barrier. In areas where heating and cooling degree days are nearly equal, a vapor barrier *is not* recommended in most localities. There are a few places, however, where the local building code may require a vapor barrier *on the outside* of the insulation. These are hot, humid climates where the contrast between the inside and outside temperatures may be nearly as great as they are in a cold climate. Check your local building code before deciding about a vapor barrier.

A vapor barrier is definitely needed on the floor of a crawl space in warm climates. Even a well-ventilated crawl space will benefit from having a vapor barrier on the ground. This not only reduces the moisture transmission from the ground, but also reduces radon gas intrusion. If a house is built on pilings, and has skirting around the perimeter, it should also have a vapor barrier on the ground under it for the same reasons. A mobile home with skirting should be handled the same way.

Attic Ventilation

Equally important in a hot climate is attic ventilation. Having a well-vented attic removes moisture from the area and helps to cool the area above the attic insulation, limiting heat transmission through the insulation which would require extra cooling in the house. Ventilation of the attic through convection, or the flow of air from soffit vents to ridge vents, is the simplest and least expensive. The formulas are the same as for ventilating an attic in a cold climate. If there is any variation, it should be on the side of *more* vents rather than fewer.

In very hot climates, especially where a roof is not shaded by trees (the best natural coolers there are), many homeowners install a power ventilation system in the roof. This can be one using wind vanes that are turned by the wind and draw air through the attic from the soffit vents. The vanes (large globe-shaped fans) are mounted on or near the ridge of a roof. As the wind turns them, they create an air flow from the attic through themselves to the outside above the roof. Often when there is no wind blowing, the vanes will turn just from the flow of hot air rising through them from the attic.

A disadvantage of wind vanes is they work equally well in cold weather, and may cool the attic too much. One method used by many homeowners in the Southwest to prevent excessive air movement in the winter is to cover the wind vane with a plastic garbage bag. This keeps the wind from turning the vane and cuts down on air flow through the attic. If this is done, there should be other vents in the roof, so the air flow isn't stopped entirely. If there is no vapor barrier in the ceiling, moisture will still accumulate during the winter, and may be enough to damage the ceiling.

Power Vent

Another method of ventilating an attic in hot climates is using a power vent. This is an electrically powered exhaust fan placed in the gable of a roof or in a housing built into the ridge of the roof. This provides a steady air flow through the attic. In larger houses, two fans, one in each gable, may be used. Fans can be controlled by an on/off switch or a thermostatically controlled unit mounted in the attic and set to turn them on when the temperature there reaches a given point.

To reduce air flow during the winter, some power vent units have gravity operated louvres, which open only when the fan is running. These are one-way louvres, so the wind can't blow through them in the opposite direction. This has two benefits. It prevents the wind from blowing into the attic during the cooler months, and keeps the wind from turning the fan backwards and possibly damaging the motor.

While passive vents should have nearly the same surface area for upper and lower vents to assure an unrestricted flow between them, take special care to have large enough intake vents when the building is equipped with any type of power venting unit, be it wind or electrically powered. The output of any fan unit should be known, and enough intake vents installed to prevent the fan from creating a partial vacuum in the attic.

Don't Block Air Flow

When installing insulation in an attic for cooling, the same problems show up as when it is put in for heating. Some ways to keep the insulation from blocking air flow between the soffit and the ridge is needed. This can be the same vent/baffles mentioned earlier, or the same construction methods described to keep the attic insulation from touching the roof sheathing and blocking air flow.

The best attic vents are those that form a continuous vent in the soffits and the ridge of a roof. Vent/baffles should be placed between every pair of rafters, the same as for a house where heating is a primary consideration. An evenly distributed flow of air is of primary importance for cooling an attic, especially where a powered exhaust is used. (Note: Only use power on the exhaust side. Do not use an electrically operated fan to force air into an attic.)

Radiant Barrier

A material that has gained prominence in the last few years is called a radiant barrier. This is a thin sheet of reflective material placed in an attic to reflect the heat from the roof back toward the outside and keep it from penetrating the living area. It is usually some type of aluminum foil, or aluminum foil bonded to a plastic sheeting. One side is usually shiny and the other dull. Unlike polyethylene vapor barrier material, a radiant barrier is highly permeable, so moisture in the atmosphere readily passes through it.

There are many brands of radiant barriers. Claims of R-value range from 7 to as high as 40 for a single layer of the material. Some are made similar to bubble wrap, having two layers with air spaces between them. Because of these claims and some of the marketing techniques used by some companies, Florida, Texas, and the United States government have taken legal action against several manufacturers of the material. Testing of some radiant barriers by the Oak Ridge Labs and the Florida Solar Energy Center, plus cold weather tests by the Canadian government do show some benefit from a radiant barrier, but also disprove many of the claims put forth by the manufacturers.

Installation

There seem to be as many methods of installing radiant barriers as there are products. One method found to be totally ineffective was installing the material directly under siding or roofing, or tight under the roof sheathing.

This eliminates any reflective value of the material. This also applies to rigid insulation with an aluminum foil backing that is used for sub-siding on a house.

Only slightly better is installing the material under roof sheathing and on top of the rafters, and letting it sag between the rafters. There must be room for air to pass between the reflective material and the roof. If it is against any part of the roof sheathing, it loses nearly all its reflective qualities.

Installation Problems

Many manufacturers of the material tell the user to lay the material on top of the attic insulation. They claim this reflects heat away from the ceiling and the insulation if the shiny side of the material is facing up, towards the roof. This does work for a while, but there are a couple of serious drawbacks. One is that dust settles on the radiant material, greatly reducing its reflective ability. In some climates problems with moisture collecting in the insulation were also noted.

Another problem with putting the reflective material directly on top of the attic insulation was found by Canadian researchers. When the outside temperature dropped below freezing, areas of an attic that had a radiant barrier laid directly on top of the insulation lost heat faster than areas without the material.

Put It Under Rafters

The best place to put a reflective barrier, according to most research reports, is on the underside of the rafters, with the shiny side down, toward the attic insulation. Leave a gap of a foot or more at the top and bottom of the material to allow an unrestricted flow of air between the roof sheathing and the radiant barrier. As with any other material used in an attic, ventilation was found to be of the greatest importance.

Some researchers expressed concern that a radiant barrier might raise roof temperatures high enough to damage some types of roofing. This was not enough to do any damage, though. If the roof is a light color, it was noted that the temperature was slightly reduced, so you might consider either light colored shingles, or an aluminum or galvanized steel roof to reflect some of the sun's heat away from the building. Some of the early claims for aluminum roof material were its ability to reflect the sun's rays and reduce the summer heat in an attic. This has been picked up by the makers of radiant barriers.

25. Superinsulation Techniques

THE HOMEOWNER living in the northern tier of states, Alaska, and Canada quickly realizes that standard construction dimensions don't allow for the insulation needed in an energy-efficient house. Unless a more expensive type of insulation is used, the frame of a house must be designed to allow for the greater thickness of the needed insulation. Because framing material is usually less expensive than high R-value insulation, engineers and builders have designed several building framing methods to use less expensive insulation but still give the insulation value needed in colder areas.

One of the first problems was the amount of wood in a standard framed wall. While wood has good insulation properties compared to many other building materials, it only has about one-third the R-value per inch of fiberglass. It follows then that the less wood there is in a wall, the more insulation can be put into that wall, and the better insulated it will be.

Arkansas Framing

With this in mind, engineers at the University of Arkansas designed a framing method that allowed for less wood, but still retained the strength of a standard wall.

In standard framing, wall studs are placed at 16-inch intervals, with double or triple studs around doors and windows. The top and bottom plates (the boards on the top and bottom of the wall) are nailed to the ends of the studs, also edgewise to the wall surfaces. An additional plate is put on the top of the wall, both to tie the walls together and to support the roof trusses, which are normally spaced 24 inches apart. With 24-inch spacing for the roof trusses, only every third truss will be directly supported by a stud. The extra thickness of the top plate was therefore needed, so conventional wisdom said, to properly support the roof.

Wood Eliminated

Arkansas framing eliminates as much as one-fourth the wood normally used in a wall. Not only was more insulation added, but insulation was put between the top and bottom plates and the inside of the house, breaking the transmission of cold through the top and bottom plates and over and under windows and over doors. It might appear the design weakens the wall, but that isn't the case. Standard framing was over built, mostly because of not having a stud directly under each roof truss. In fact, it was so overbuilt that many building codes allowed the use of utility or #3 grade lumber in wall framing.

Arkansas framing uses construction or standard (#1 or #2) grade lumber, but since it uses less material, the cost tends to even out. Arkansas framing changes the way the top and bottom plates are placed, and how the headers over the windows and doors are mounted into the wall. The studs are placed on 24-inch spacing, and the top and bottom plates are turned on edge and mortised or inset into the studs. The door and window headers (the boards directly over the doors and windows) are also mortised into the studs and turned on edge, eliminating extra studs around the doors and windows and allowing for insulation to be put into the space between the headers and the inside of the wall.

Carpenters have known for a long time that one board on edge was actually stronger than two boards of the same dimension placed flat. In addition, with the studs at 24-inch intervals, each roof truss rests directly on a stud, giving the strongest possible support to the roof.

Extra Labor

The main objection to Arkansas framing, from the viewpoint of a building contractor, is the extra labor needed to cut the notches where the top and bottom plates and the window and door headers rest in the studs. But trading labor for insulation has proven to be worthwhile, as the insulation helps reduce energy bills for years to come.

Another problem is with the interior wall sheathing. Most building contractors use ½-inch gypsum board. With the studs on 24-inch centers, ½-inch gypsum board tends to sag between the studs, and over the years makes for an unsightly wall. (Many building codes prohibit the use of ½-inch gypsum board on greater than 16" stud spacing.) This problem is easily overcome by using ⅝-inch gypsum board as interior wall sheathing. Again, the reason for using ½-inch material is cost. It costs less but takes only slightly less labor to install than the ⅝-inch material. (Not that much less, as many builders are finding out.) If the electricians who install the receptacles and the carpenters who put in the doors and windows know in advance that ⅝-inch material will be used, they can plan for it, and any extra cost will be negligible.

The principal objection to Arkansas framing is found in some local

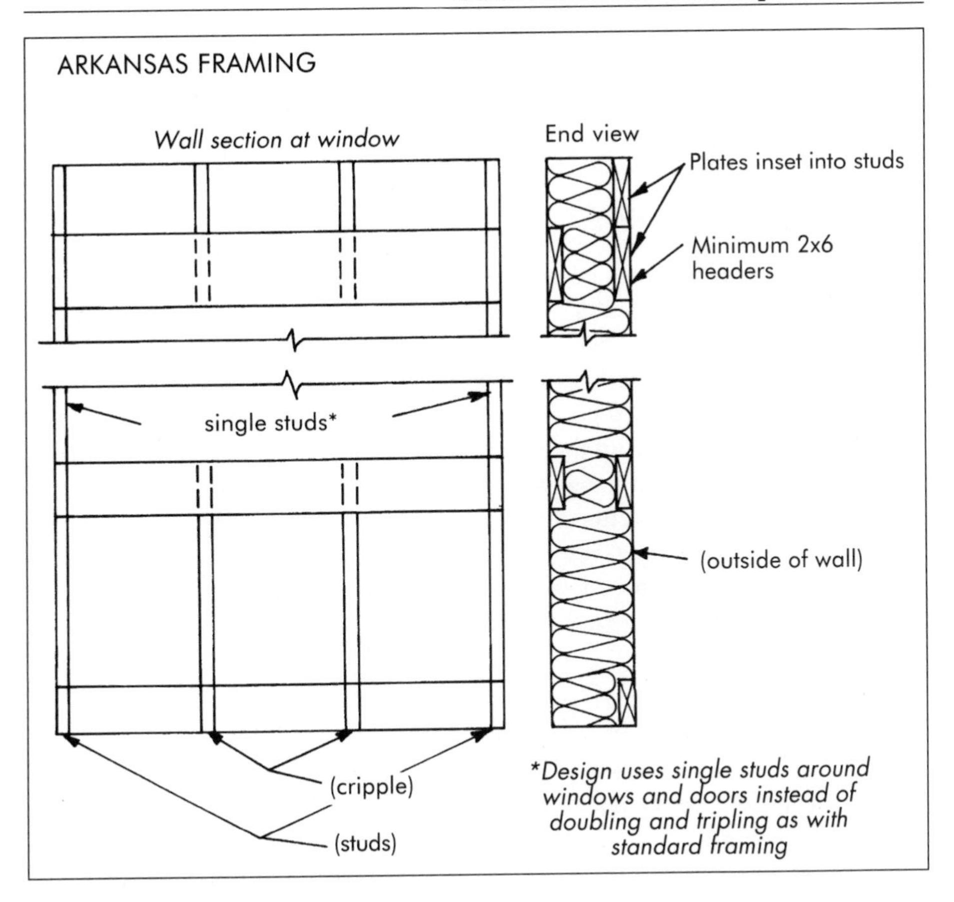

building codes. This framing is similar to the balloon framing used in the United States from the early 1800s until after 1900. Top and bottom plates and support ledger boards supporting the floors were mortised into the studs.

Fire Hazard

A lot of two- and even three-story houses were constructed this way. A wall the full height of the house could be framed, without a break at each floor. The problem came because no insulation was used, and there was an open space between each stud from attic to basement. This was fine in summer, with a well-vented attic, as cool air circulating from the basement to the attic in the walls helped cool the house. However, if a small fire started in the basement, it quickly became a large one, because the wall spaces also acted as chimneys and drew the flames upward, quickly enveloping the house. There are some claims that the rapid spread of the "great Chicago fire" was aided by this type of construction.

The name balloon framing was given in derision by old-time carpenters who built post-and-beam and log houses, and masons who built stone buildings. They claimed houses built that way would "fly away like a balloon" with the first strong wind.

Scandinavian Method

Another system of construction is found in some Scandinavian countries and may be combined with either Arkansas or standard construction to increase the amount of insulation in a wall without greatly increasing the cost.

This involves placing furring strips horizontally on the room side of the wall studs. The method used in the Scandinavian countries is to insulate the wall, then put up the vapor barrier, then place the furring strips. All wiring and plumbing that would be in the wall are put on the room side of the vapor barrier. This eliminates a hole in the vapor barrier at each outlet and switch box as well as where plumbing (not a good idea in an outside wall) goes through. The interior wall sheathing is then placed in the usual way except the nailers are horizontal instead of vertical.

Other Methods

These systems may be used in climates where approximately 6 inches (R-

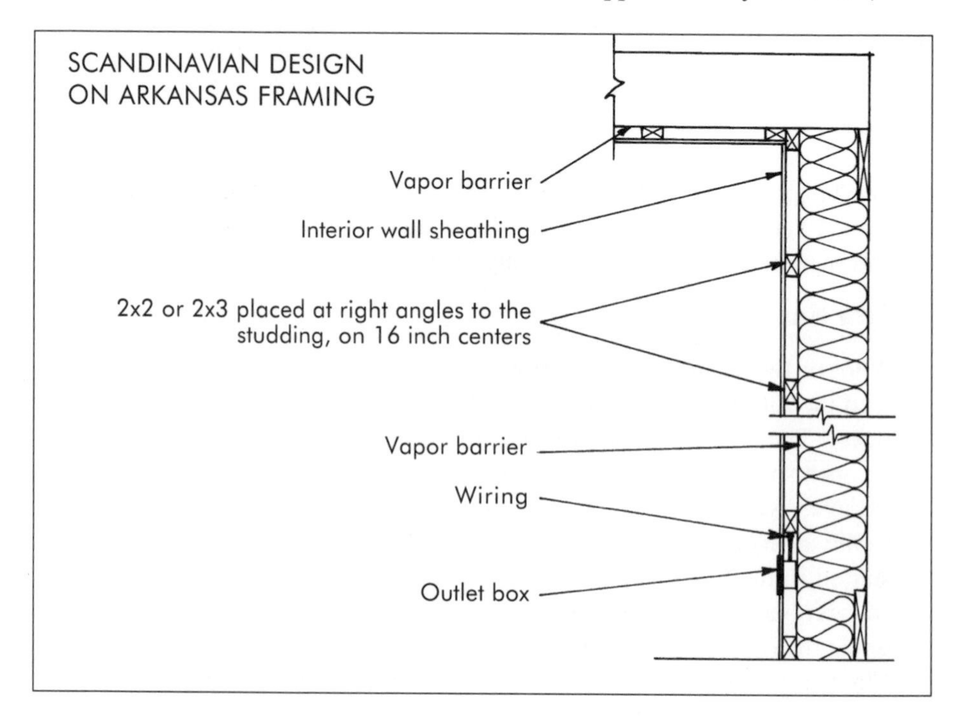

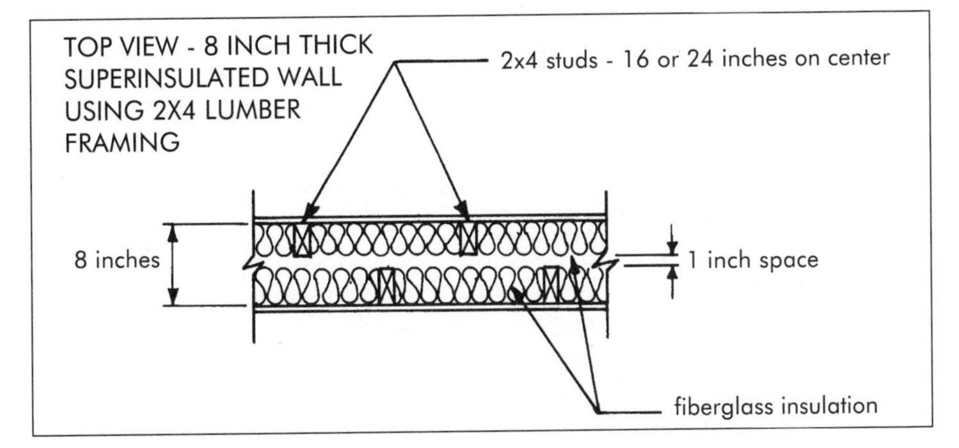

19) of fiberglass insulation will be enough in an outside wall. For colder climates, two other methods of construction are available, one using the less expensive fiberglass insulation and another using some unique properties of foam or rigid insulations.

The first of these is a double-framed wall. Instead of one wall, either standard or Arkansas framing, two are built, one from 1 to several inches inside the outside wall. Both walls are insulated with whatever thickness of insulation will fit them, and insulation is also placed in the space between the walls. This is the system most often used in "superinsulated" houses. It breaks any transmission of cold through the studs and allows the builder to increase the insulation in a wall to whatever thickness is desired, simply by placing the inner wall at whatever distance is needed from the outer one.

The walls are constructed from either 2x4 or 2x3 dimension lumber, which is less expensive than building a wall out of lumber the dimension of the wall. In this area at the present time, an 8-inch thick wall built from two 2x4 walls costs approximately a third *less* than an 8-inch wall made from 2x8 lumber. An added savings is there are two walls supporting the roof, so if allowed by code the studs of both walls may be on 24-inch spacing. Window and door frames need to be ordered to fit the thicker walls. Plan ahead.

Sandwich Wall

A new method of construction, where manufacturers are set up to build it, is the sandwich wall. This is constructed of plywood or flake board sheathing glued to polystyrene, polyurethane, or isocyanurate rigid insulation. A wall of any reasonable length or height may be manufactured, then hauled to the building site and put in place on a prepared floor. The wall can be made whatever thickness is desired for the necessary insulation value.

A variation of this is a wall having studs and framing made from dimension lumber, with foam insulation forced into the cavities through holes in the bottom of the wall. Either may have doors and windows roughed in, as in regular construction, or they may be cut in at the site, after the walls are set up.

The advantage of this building system is how fast a building may be put up. This reduces construction labor costs considerably. Where inclement weather may hinder construction, closing in a building in a day instead of a week is a factor to be considered.

Disadvantages are the cost, which may be much higher than standard, and limits on just how thick the walls may be made, limiting the insulation value of the wall. Local building codes may also limit use of sandwich walls. Many local codes specify what is allowed. If a particular type of construction is not specifically allowed, it is assumed to be prohibited. No matter that it is better, it won't be allowed.

Another item to consider is the claimed R-value of the walls. The insulation used is either polystyrene, polyurethane, isocyanurate, or polyurethane foam. If the claims for the wall exceed the numbers in the chart of R-values in this book, perhaps the manufacturer is overstating a bit. If the wall makers can't or won't give you the particulars on the insulation they use, perhaps it would be better to select another form of construction.

Use Treated Lumber

Building basement walls or the foundation of treated lumber (all-weather wood) has been mentioned earlier in this book. It can also be considered as another construction method to enhance the use of insulation. It is legal under most building codes and government loan programs. The spacing of the studs, how it is placed on the footing, and the dimensions of the material may differ from locality to locality. It is a good way to get the most insulation for the least cost in either a basement or a crawl space. Information on all-weather wood foundations may be obtained from your local university Extension office or from the U.S. Forest Service, Forest Products Laboratory. Information is also available from many building supply firms that sell the treated lumber. (Address of the Forest Products Lab is given in the appendix.)

Insulating Doors

No matter how well you insulate the walls, you still have to get in and out of the house. Along with the heat loss every time a door is opened, doors are also a source of air leakage. The addition of an Arctic entry, commonly called a mud room in the north country, will eliminate these problems and reduce energy bills by keeping the heat where it belongs, in the house. The Arctic entry is a small room built around each outside door, either inside or outside. It needs to be large enough so the outside door is closed before the inside one is opened. This eliminates that cold blast of air whenever a door is opened to the outside.

Putting a source of heat in the entry also helps. With a closet to hold coats and boots, and a place to sit when removing boots, it will eliminate a lot of tracking into the main part of the house.

New methods of construction or adapting old ones to enhance the use of insulation is going on continually. Most are good ideas, but are still subject to the whims of local building codes. If you want to try some new type of construction, check with your friendly building inspector first. Also keep in mind the old adage, "If it sounds too good to be true, it probably is."

26. Ways to Improve Insulation

MANY ITEMS ASSOCIATED with comfort in a house seem to have little to do with insulation. But to get the full benefits of insulation, they must be considered as part of the insulation job. Placement of windows, air circulation, and air exchange are three such items to consider if your house is to be comfortable. This especially applies to a well-insulated house.

Size, design, and materials of windows were covered in a previous chapter. There is another facet, however, to windows. This is what building engineers call passive solar heating. This means using the sun to provide part of the heat. Properly located windows become a part of the heating system and can contribute many dollars worth of heat to a home.

In a properly insulated house, the sun shining through a south-facing picture window may provide all the heat needed for the day if there is an air circulating system available to distribute the heat throughout the building. Without a means of air circulation, the room with the window can become unbearably hot while the rest of the house gets colder and colder. A reason for this is that the thermostat is in the main living area of a house, which is often near such large windows. If the thermostat is warmed by the sun, it will not call for heat, and the blower in the heating system won't run.

For an older heating system, with a simple "heat on demand" thermostat, it is easy to add to or change the controls to either incorporate a simple fan switch, which allows the fan to run all the time or that activates the fan if the area where the thermostat is located gets too warm as well as too cold. Most new forced air heating systems have part or all of these extra controls.

Air Circulation

Many heating systems don't have powered air circulation. Some depend on natural forces such as convection and radiation to put heat in a house. These utilize the "well-known" fact "warm air rises and cold air falls." Out in the open this is very true, but in a tightly enclosed house things don't seem to work that simply.

In a closed room, where there is nothing acting to move the air around other than convection, the air tends to form layers or stratify, and stop moving. The air at the ceiling may reach 90-100° while the air at the floor is closer to 55°. Since we do most of our living in the bottom 5 feet of a room, either the heat must be turned up or we, and especially our feet, will be cold.

Just a small electric fan, placed near the ceiling, will make a big difference in both the comfort and the cost of heating that room.

More elaborate, though inexpensive, air circulation systems may be utilized. If the ceiling is high enough for safety, a large cafe fan may be used. Operating at its slowest speed, where the movement of the air isn't noticeable to the occupants, it will make a room more comfortable.

Another simple device which requires no additional electrical wiring is a hassock fan. This is a fan mounted in a hassock that sits in any convenient place on the room floor. It isn't as efficient as a cafe fan, but doesn't cost as much, either.

There is also a small unit available that is designed to take warm air from the ceiling and blow it across the floor. It looks like a length of plastic sewer pipe with a small fan in the base. It is most often located in a corner, out of the way of the users of the room.

Stale Air

With houses being built tighter and tighter, air in a house will become stale and fouled by pollutants, some of which is caused by the occupants breathing.

Older houses had enough leakage through the walls and around the doors and windows to produce what building engineers call an air exchange every few hours. Sealing the walls with plastic film and the windows and doors with weather stripping and caulking has stopped this natural air exchange. So the occupants will be healthy as well as comfortable, some means of changing the polluted air for fresh air from outside should be part of the heating system.

Add Ducts

If the house has a forced air heating system, it is easy to add ducting to introduce fresh air into the system. This is usually placed on the return air side of the furnace, where the cold air will be drawn in and mixed with the inside air before it is heated in the furnace. This eliminates cold drafts from the heating system. Either manually or electronically controlled shutters may be installed at the air intake to regulate the amount of cold air brought into the house at any given outside temperature.

In a house heated by a convection and/or radiation system, the problem is more complex. The major problem is introducing cold air into a warm house. If it is brought in directly from the outside, the area near the entry point will be uncomfortable. There may also be moisture condensation near the entry, which may damage the interior of the house. To counter these problems, some means of tempering or partially heating the incoming air is needed. In the chapter on radon gas, a unit was mentioned that does this by using the warm air it exhausts from the house to partially heat the fresh air it draws in.

There are a variety of these units on the market. Some use exhausted air for heating the incoming air. Others use some type of a heating unit, either independent of or part of the home heating system. An example of this is an air exchange unit having a small radiation unit plumbed into a hot water heating system. Another uses an electric heater built into the exchange unit. These add a little to the home energy costs, but for the health and comfort of the occupants they are worth it.

Passive Airflow

Another air exchange system, using no energy, only simple ducting, has been designed by Canadian building engineers. They call it a passive airflow system. It relies on the exhaust devices in a house to draw fresh air into the house. They are the clothes dryer, the kitchen and bathroom exhaust fans, and the chimneys for fireplaces or wood stoves.

This is also called a make up air system. Don't confuse it with the combustion air system required for gas, oil, wood, or coal heating systems. The passive system replaces the air exhausted by a dryer or other device, while the powered system actively removes and replaces the air in a house.

It has been calculated that a minimum of 66 feet of ducting is required to properly temper the air drawn into a house. This includes a duct hood at the entry point, and one or more 90° elbows. It has also been calculated that a flow of 200-250 CFM (cubic feet per minute) is necessary for most houses, so a duct size of 6 to 10 inches may be needed.

Determining Duct Size for Passive Airflow System

CFM	Duct Size
15	3 Inches Diameter
29	4 Inches Diameter
59	5 Inches Diameter
95	6 Inches Diameter
148	7 Inches Diameter
212	8 Inches Diameter
296	9 Inches Diameter
381	10 Inches Diameter

1. Ascertain the exhaust airflow in CFM (cubic feet per minute) of each appliance. (Range hood, Dryer, Bath fans, etc.)
2. Add the above figures together.
3. Add 75-100 CFM for a fireplace or wood stove.
4. Find total CFM figure on table, read across and find duct size.

This chart is for a duct length of 66 feet, having a duct hood and one or two 90° elbows and a pressure differential of 15 pascals.

How can 66 feet of 6-inch or larger ducting be put in the average house? In a house with an insulated crawl space it can be put there, with an opening into the house located where it won't cause a cold draft. The ducting is out of the way, and the warm air in the crawl space will temper the incoming air. For a house with a basement it would be better to use the powered air exchange unit.

Combustion Air

The combustion air mentioned in a preceding paragraph is what a fired heating system requires for safe and complete burning of the fuel used. All building codes call for a separate combustion air inlet for oil- and gas-fired appliances. In addition, many codes require such appliances be located in a separate room, blocked off from the rest of the house by a "fire rated" door. Two combustion air inlets are usually required; one to be within 12 inches of the floor and one within 12 inches of the ceiling. A total size of 1 square inch for each 4000 Btu of heating is stated by some codes. More than that is better.

Here again there can be a problem with introducing freezing air into an area with plumbing. Baffles placed to divert the air from any exposed plumbing need to be installed, or some way of drawing the air indirectly, such as through an insulated crawl space, may be used.

Duct to Outside

Another way of obtaining combustion air is available. This is called a closed combustion system, and comes as part of some new heating systems. The appliance draws its own combustion air through a separate duct directly from the outside of the house. This eliminates any openings into the house that could cause frozen plumbing and condensation problems.

In some cold areas, and where allowed by the building code, powered shutters are used on combustion air intakes to prevent freezing problems. These are wired to the heating system controls so they open when the heating unit is about to fire, and close when no combustion air is needed. They are mostly used in commercial systems for large furnaces (1,000,000 Btu and larger). Some home building codes prohibit these.

For a house with a wood stove or fireplace, make up air and combustion air are often the same. In a tight house, a lack of incoming air can cause a smoking stove or fireplace or one that will not burn well. Manufacturers of some new fireplace units and woodstoves are adding combustion air ducting (closed combustion) so combustion air may be brought from outside the house and eliminate these problems. These should have some way to cut off the air flow when the unit isn't being used to prevent cold from entering the house.

27. Insulation Pays for Itself

SEVERAL TIMES IN THIS BOOK I have stated that insulation pays for itself. Since the addition of insulation to a house decreases the amount of energy needed to heat or cool that house, this should be obvious. The next question is also obvious: How much will be saved, and how long will it take to pay for the insulation?

Building engineers can turn to their computers and quickly come up with some figures. For a new house, the Canadian "Hot 2000" program can break the answer down to project savings for each type of fuel that could be used. For older houses, computer projections can show savings for the type and amount of insulation used. One question leads to another: How accurate are the computer projections?

A study in central Illinois, which tracked twelve houses for a period of nine years, is enlightening. These older houses had been retrofitted with insulation in an attempt to make them more energy efficient. Computer projections were done prior to the re-insulating, then the energy uses of the houses were compared to the projections. The computer projections were found to be an average of 10 percent off the mark.

The actual figures, though, were not discouraging. The average reduction in energy consumption was 23 percent, with a low of 6 percent and a high of 43 percent.

Payoff Time Varied

The time required to pay for the insulation from the savings in energy costs, (payback time) also varied widely. The shortest payback time was four years. The longest (projected, since the study didn't last that long) was twenty-seven years.

Actual cash savings, figuring in a federal income tax credit, was from $55 to $320 a year, with an average of $140.

Another study, based in the upper Middlewest but tracking houses in different parts of the country, approached the question another way. The researchers looked for the savings from the addition of specific R-values of insulation in the walls and ceilings of older houses.

The first portion of the study covered houses having no insulation in the attic. R-19 insulation was installed, which reduced heating costs by 13 to 21 percent. Increasing the insulation from R-11 to R-30 reduced heating costs by an average of 13 percent.

Definite Savings

Both the studies show a definite savings from putting in or adding insulation. They also show a wide variation between individual houses, even when they have the same amounts of insulation added. A small reason for the variation in the first study could be attributed to different house designs. A similar reason could be set forth for the second study; different houses in different parts of the country.

The Human Factor

Another reason, and a major one, quickly comes to mind to those of us who have been involved in the building trades for a number of years. That is the human factor. The occupants of a house have as much to do with the energy consumption of that house as the design. If they are energy-conscious, they will take advantage of every energy-saving device and method they can afford. If they aren't especially energy conscious . . .

Changes in the occupancy can vary from different families (some statistics claim the average family moves every five years, so there could have been several such changes in a nine-year study) to the children growing up and moving away, or children being added to a family.

A growing family alone will cause a lot of variation in energy consumption from year to year. Small children spend a lot of time playing on the floor, so more heat is required to keep the floor area warmer. Pre-teen children spend a lot of time coming and going, opening and closing outside doors — and often leaving one open. High school and college-age children are not in the house as much as when they were younger, which can cut energy costs drastically.

There are other variables over which the homeowner has little or no control that can affect the cost of heating a house. The weather, or climatic variations, can make a large year-to-year difference in heating and cooling costs. This is another reason for tracking costs for a period of years before concluding what the savings may be from a specific insulation project.

Trees Can Cut Costs

Other outside factors could be the growth of trees that shade the house in the summer and reduce cooling costs, or the growth of a windbreak that will protect the house from cold winter winds. Additional houses constructed in the area can also affect energy costs for a home. Building up an area can raise the average temperature of the whole area, for instance. Other buildings may also act as a windbreak, and reduce the amount of energy needed for heating a nearby house.

Another human factor that often comes into play is a tendency by many people to move more and more into energy conservation once they have started. Weather stripping doors and windows to cut down on drafts, for instance, can lead to adding attic insulation, wall insulation, and Arctic entries as each reduces energy costs. This may also lead to some life style changes, to gain further reductions in expenses. For many people who "could care less," all it takes is a start. From then on, it's like a row of dominoes.

All the foregoing may lead someone to believe that any payback from adding insulation and other energy-reducing items will be a long ways down the road. Total payback may be several years off. There will usually be an instant payback, though, both tangible and intangible. The tangible one will be an immediate reduction in energy use. Actual costs, because of the fluctuating cost of fuel oil for instance, may or may not drop.

An intangible benefit may be something as simple as a lower thermostat setting keeping the house at the comfort level that the occupants are used to. A lower thermostat setting puts less demand on the heating system, therefore reducing energy costs.

Estimating immediate payback versus a long-term payback is something that can get the buyer or owner-builder of a new house in trouble. Some architects, builders, and insulators will not put the optimum amount of insulation in a new house, arguing that the payback curve flattens quickly and extra insulation won't pay for itself.

There may be a point where a certain amount of insulation could be considered too much, but the formula given in this book will not do that. It was arrived at after research indicated heat loss could only be reduced so far. The formula will give the value of insulation needed to attain that minimum heat loss, no matter what the climate, because the formula is based on the degree days for that specific climatic area.

The reason for the builder of a house to not put in the maximum amount of insulation is the same as for putting it in. Financial. Extra insulation translates into extra cost for a builder, both for material and labor. That extra cost may cause him to lose the bid or the sale. Therefore it is better to skimp on insulation.

Most building codes don't specify insulation. A code that specifies a minimum amount of insulation aids this deception. Remember, building codes are minimum specifications. A builder who "built it according to code" may be getting by as cheaply as is legally possible.

Appendix

For technical information about insulation and other energy problems in a house

NATAS (The National Appropriate Technology Assistance Service)
U.S. Dept. of Energy
P.O. Box 2525
Butte, MT 59702-2525

1-800-428-2525, 9 a.m.-4 p.m. (Mountain Time) weekdays.
In Montana, 1-800-428-1718.

Oak Ridge National Laboratory
Oak Ridge, TN 37831

Insulation planning computer software

ZIP 1.0 software and program information:

MTS Software
3534 Knollstone
St. Louis, MO 63135 ($5 charge)

National Technical Information Service
Springfield, VA 22161
(703) 487-4600

(Give the first three digits of your Postal ZIP code to obtain the correct program for your climatic region.)

HOT 2000:

HOT 2000 Sales
Canadian Homebuilders Assn., Suite 702
200 Elgin St.
Ottawa, Canada K2P 1L5

Model insulation codes and other information

Alaska Craftsman Home Program, Inc.
P.O. Box 876130
Wasilla, AK 99687

American Society of Heating, Refrigeration, and Air Conditioning Engineers
1791 Tullie Cir. NE
Atlanta, GA 30329

Moisture problems and home energy conservation booklet

Energy Administration Clearinghouse, Michigan Dept. of Commerce
P.O. Box 30228
Lansing, MI 48409
1-800-292-4784, 9 a.m.-4 p.m. (Central time) weekdays

Radiant barriers and warm climate vapor barriers

Florida Solar Energy Center
300 State Rd. #401
Cape Canaveral, FL 32920

(305) 783-0300

Oak Ridge National Laboratory
Oak Ridge, TN 37831

To Obtain Information about All-Weather Wood Foundations and Basements

Forest Products Laboratory
1 Gifford Pinchot Dr.
Madison, WI 53705-2398

To Obtain a Homeowner's Glossary of Building Terms

R. Woods
Consumer Information Center
P.O. Box 100
Pueblo, CO 81002

(Publication #134X, $1 fee.)

Glossary of Terms found in this book

FOOTING. The spreading course found at the base of a foundation or basement wall.

FOUNDATION. The supporting portion of a structure below the first-floor construction. Includes the footing.

FRAMING. The timber structure of a building which gives it shape and strength; includes interior and exterior walls, floor, roof and ceilings.

FURRING. Narrow strips of wood spaced to form a nailing base for another surface. Furring is used to level, to form an air space between two surfaces, and to give a thicker appearance to the base surface.

INSULATION. (Thermal) Any material high in resistance to heat transmission that is placed in structures to reduce the rate of heat flow.

JOIST. One of a series of parallel framing members used to support floor and ceiling loads, and supported in turn by larger beams, girders, or bearing walls.

PURLINS. Framing members laid at right angles to rafters, joists, or studs to support roofing, wall surfaces, ceilings, or flooring, to gain space for insulation or ventilation. (Larger than furring strips.)

RAFTER. One of a series of parallel framing members used to support the roof of a building. The rafters of a flat roof are also sometimes called the roof joists.

STUDS. A series or parallel vertical wood or metal framing members in walls and partitions.

TRUSSES. A structural unit designed to provide rigid support over a wide span. If used to support a roof, the top chord corresponds to rafters, and the bottom chord corresponds to the ceiling joist.

VAPOR BARRIER. An impermeable material usually placed on the warm side of an exterior wall to prevent the moisture in the air of a house from migrating into the walls of the house.

Index

Page numbers in italics indicate charts, illustrations or tables.